Wild Arum

The Secret Life of
Lords and Ladies

By
Lynden Swift

Green Yaffle Press

© Green Yaffle Press
Concise Edition
2017
Bristol

E: info@greenyafflepress.com

W: www.greenyafflepress.com

978-0-9575829-5-8

Dedicated to everyone who helped to make this book a reality.
Thank you.

Acknowledgements

A great many people make a book like this possible so firstly thank you to all those who answered my letters and emails, who assisted with my queries and were patient with my poor academic knowledge. Particular thanks must go to the following:

Morgaine, for invaluable input with proof reading and editing.

Malmesbury Abbey Gardens, for letting me sneak in before official opening times to photograph their particularly stylish Arums.

Fergus Drennan, for kind permission to use his photographs and recipes.

Marcus Harrison, for information and assistance with researching the history of Arum's culinary usage.

The staff of Bristol Library, for ordering ever-more obscure journals and ancient books from far and wide.

Sarah Melamed, for kind permission to reproduce her recipe for Arum soup.

Robert Johnson of the Poison Garden website and Jill Turner of the Royal Botanic Gardens at Kew, for assistance with information on Arum poisoning.

Georgia Glover, on behalf of the Estate of Geoffrey Grigson, for kind permission to use the information from his book 'An Englishman's Flora', for Grigson's sources for Arum's local names.

Roberta Rostirolla of Lucente for kind permission to use the images of the Arum Lamp and for information on the background of their creation. Richard Mabey, for assistance with information on the background to the carving of Arum in the Four Elms church.

CONTENTS

PREFACE

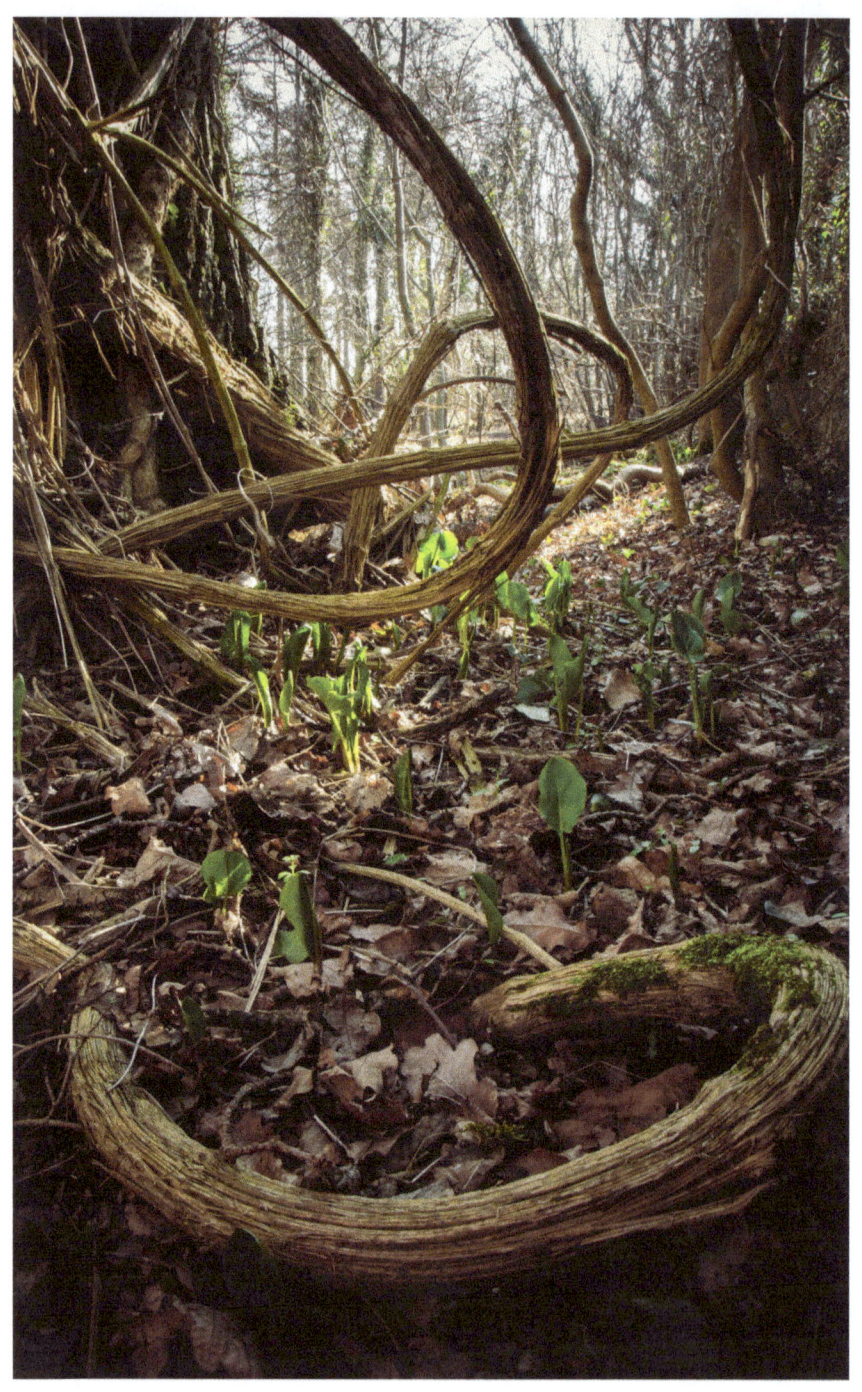

W hy a book on Arum? Let's take a walk through the British countryside in the early spring or late summer. Cast your eye under the hedge-bank or into the shade of the woodland you pass. Do you see any strange, hooded creatures in the undergrowth or perhaps a procession of tall stems offering up inviting yet threatening clusters of bright red berries? If so, you have stumbled upon one of the most distinctive and unusual plants we have with us in this country.

It was on such springtime walks that I became enamoured with the Wild Arum and fell under its spell. Its distinctive appearance, a family trait, makes Arum a wonderful plant to photograph and initially I viewed the plant simply as an unusually stylish photographic subject. This superficial phase was not to last long. What began as a quick query to learn a smattering of background unearthed such a rich and interesting history that I had to find out the whole story. The result is this book you are now reading. In many ways, it's a homage, a love letter even, to Arum.

If you wish, you can read this book sequentially – from this point on through to the last page. Alternatively, you can jump in and skip about as you see fit. There's very little which builds on that which has gone before and what there is I'm sure you'll cope with. So take your pick and enjoy reading about this fascinating plant. The photographs do follow a sequence of a sort, in that they have been arranged to follow the plant's cycle through the year, from its first appearance in spring to its red berry finale of late summer. But, as with the written sections, your enjoyment will not be lessened by taking a peek at the summer pictures before you have experienced the spring.

This is not an academic or formal monograph of this plant; on the contrary, it is a celebration in pictures and words of the most interesting and stylish-looking plant we have in our British countryside. I take liberties with contemporary taxonomy and give free rein to imagination when trying to discover and express the inner 'soul' of the Arum plant. If this were a dinner party I would be exuberantly introducing you to a great friend of mine and painting them in the best of lights. In that spirit, I hope you enjoy reading and exploring this book as much as I have writing it. If you have decided to begin at the beginning, then that's just great. We begin, as with any introductions, with names. Let's meet Arum.

On the Sight of Spring

How sweet it used to be, when April first
Unclosed the arum leaves, and into view
Its ear-like spindling flowers their cases burst,

Betinged with yellowish, white, or purplish hue!
Ah, how delighted, humming on the time
Some nameless song or tale, I sought the flowers!

Some rushydyke to jump or bank to climb
Ere I obtain'd them; while from hasty showers
Oft under trees we nestled in a ring,
Culling our lords and ladies. O ye hours!
I never see the broad-leaved arum spring,
Stained with spots of jet, I never see
Those dear delights which April still does being,

But memory's tongue repeats it all to me.

John Clare.

Wild Arum

The Secret Life
of
Lords and Ladies

Lynden Swift

INTRODUCTION

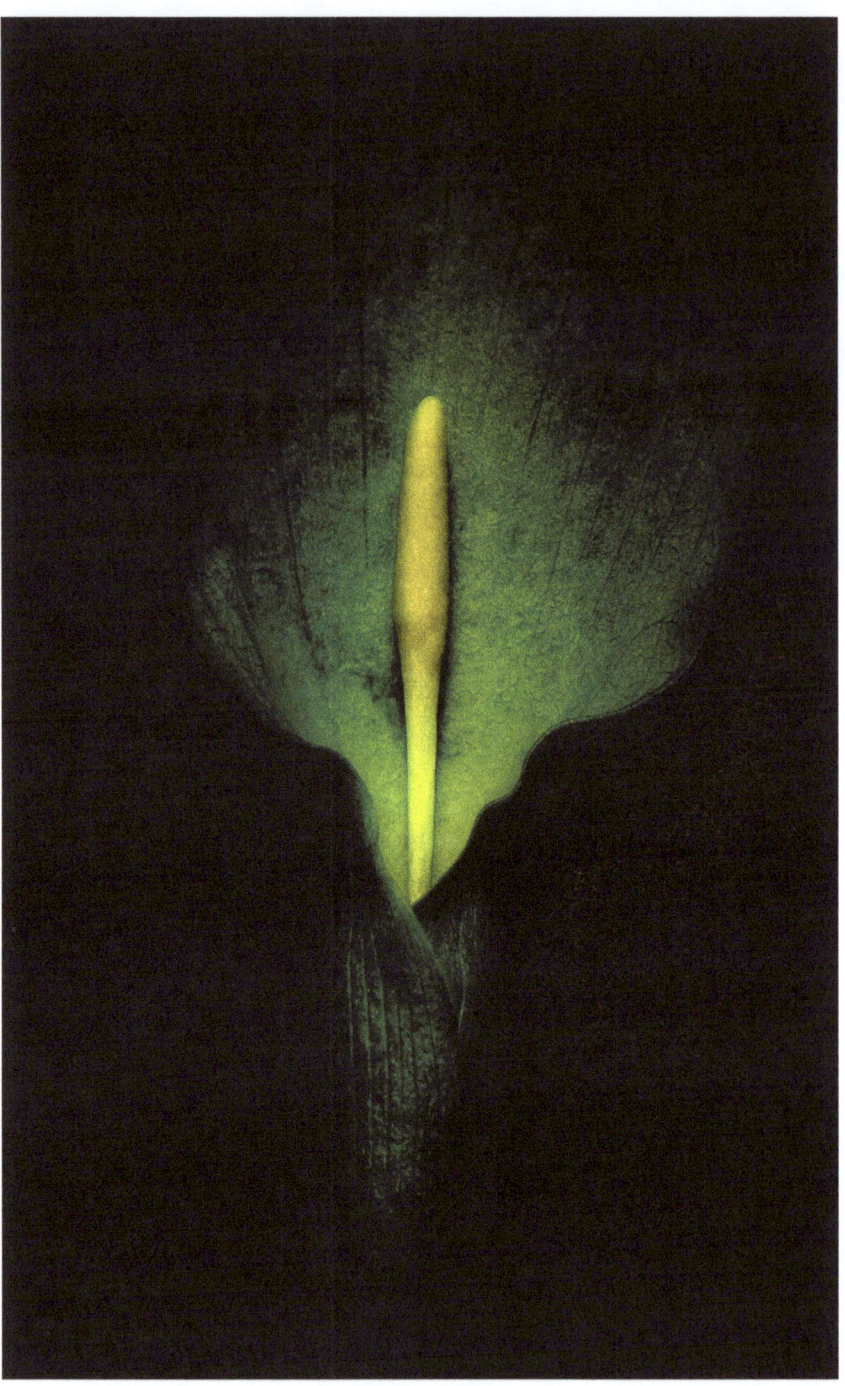

0

'This crafty and malignant antediluvian vegetable'

T hat is how Lords and Ladies was described in 1899, when it was
believed that its red berries had evolved to poison birds that ate of
them to become "huge manure heaps for the growth of the young plant".
A crafty and malignant plant indeed.

For many, Arum still carries a certain air of danger or mystery.
From its sexually suggestive cowls which announce its springtime presence
to its maces of lipstick-red berries which perch atop slenderously erect
green stalks in late summer, it is not surprising that the 'Gentleman's
Finger' (as it was salaciously known in Wiltshire) has garnered more
common names than any other UK plant: well over 100 at the last count.

To know the Wild Arum is to steep oneself in the history of our
ongoing relationship with plants, to delve into the very roots of herbalism
and botany and to take in the grand view of the long evolution of science
and medicine. It is to go on a journey which begins in ancient Greece
with the earliest herbal manuscripts ever written and ends with the latest
genetic research into the evolution of plants. It is to step into the tradition
of writing and publishing, taking in along the way the first herbal guides,
much myth and folklore, Elizabethan high fashion, rampant plagiarism
and hibernating bears. It's quite a story.

Over the centuries the Wild Arum has been used as a food, a
medicine, a fashion accessory, a tool to get rid of unwelcome guests and a
symbol of sexual intercourse in medieval art. The Egyptians carved it on
the walls of their temples and it shares its name with their ultimate creator
god. It was mentioned by Theophrastus, student of Aristotle, 200 years
before the birth of Christ and featured in the first ever encyclopaedia as
written by Pliny the Elder in AD77. It resuscitates bears from hibernation,
can raise its temperature to an astonishing 32°C and is said to glow at
night, giving it the name of fairy lamp.

With Arum, one is never far from the potent life force of sexual
creativity itself. Arum has symbolic links with serpents, with fire, with

life, death and resurrection and with the interplay of all these things. Be on your guard then, for in getting to know Arum, there's no escaping the profundity of its inner nature.

CHAPTER 1
OF FIRE & SERPENTS

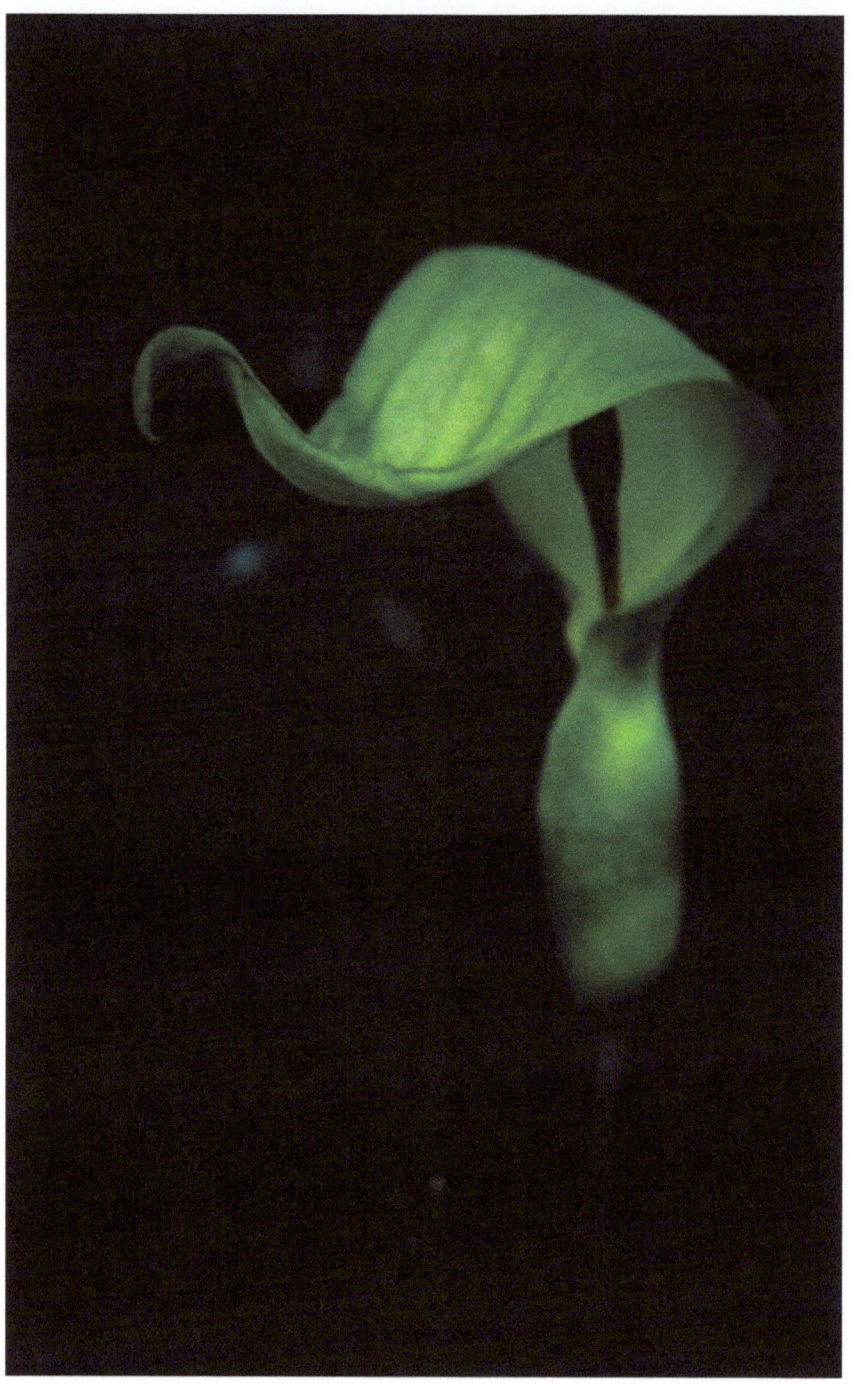

1

F or most plants, a single name is all they have left. Those more characterful have successfully clung on to a handful of so called 'country names', echoing old superstitions or now forgotten historic uses. For the majority, their names are now frozen; defined and recorded and finalised. Our linguistic creativity in simply making up names for the plants around us has withered away, disapproved of by 'official' guide books showing the sole and true 'correct name'. People might even say that they do not know what a plant is, because they do not know its 'correct name', rather than simply making one up on the spot in a fit of herbaceous creativity. So much the poorer are we then. But not with the Arum.

Due to its distinctive shape and ever-sexually-suggestive pull on our imagination, the Wild Arum has inspired a wildly excessive number of folk names: 165 at the last count in English alone, quite possibly more than any other UK plant.

Many are startlingly similar even when hailing from different parts of the country, such is its archetypal power over us. They are the perfect antidote to linguistic paucity and good mannered restraint. So next time you come upon the Wild Arum whilst exploring the spring-time woods, what can you do but think up a new name to add to the list? For now, let's take a look at what Arum has already stirred in our imaginations.

Typically for this plant, even its scientific name is steeped in mythology and mysticism. The botanical name *Arum maculatum* first appears as far back as 1588, when Tabernaemontanus used it in his great herbal: the *Neuwe Kreuterbuch*, the illustrations of which went on to be used in Gerard's herbal of 1597. At that time, the name was used in a fairly loose fashion and didn't necessarily refer specifically to our modern *Arum maculatum*. The plant was more commonly known, when it was specifically named at all, as *Arum officinarum* – or at least it was so called by Matthaeus Lobelius, the French botanist who began the first concerted attempt at botanical classification during the 1500's. In reality, the name didn't stick, but it does tell us that it was already the 'type plant' for the Arum genera: the reference or starting point to which all other varieties were compared.

The name and the plant as we know them did not come together until 1753, when Linnaeus formally named the plant as such in his famous *Species Plantarum*, which defined the format for scientific classification that is still used today.

Prior to this, 'arum' was the generic term used by the ancient herbalists to describe many different but visually similar plants, including some now no longer classified as part of the Araceae family. It is an interestingly appropriate name for this group of plants. The word 'arum' itself is usually taken to be the anglicised version of the Greek *aron*, which can often be seen in the ancient herbals from Dioscorides onwards, sitting conspicuously amongst a rabble of more ribald local names. It's often suggested that *aron* is derived from the Arabic word *ar* meaning 'fire', due to the caustic taste of the leaves of arum plants. This link is often attributed to Cecil Prime but the earliest reference is in Anne Pratt's book of 1855, where she writes of a 'Professor Hooker' who has linked the 'ar' in arum with the *ar* of Arabic and the *aur* of Hebrew. Professor Hooker was onto something.

Ar is far older than Arabic – it is an ancient root word linked to the element of fire in the sacred sense. It makes its first appearance in early Sumerian and Egyptian records and has retained its original association with fire in many languages and words since. In ancient Etruscan it is the word for divine fire and in Egyptian it is related to the word for sacred light. Light and fire are often mythologically and linguistically linked, so it is possible that the name for this plant hails from the very deepest roots of Western culture and is linked to the sacred rather than the mundane qualities of the 'fire/light' element. Already, Arum is telling us it is no ordinary plant.

Arum is also one of the lesser used renditions of Atum, the old Egyptian god of creation. He was later merged with Ra, who symbolised the Sun, which of course is linked to both light and fire. Arum (the God) is often depicted as a serpent. Arum is also the Hebrew word for serpent (notably the serpent in the Garden of Eden), but it also carries the linked meaning of wisdom and cleverness and Arum the creator God was obviously no slouch when it came to creativity. The shared name is undoubtedly a coincidence, yet the plant Arum has long carried a strong association with serpents and even contains poisons which are similar to those in snake venom. Arum – scratch the surface and it's a long and beguiling linguistic mine that beckons.

Whatever the serpentine route its scientific nomenclature has taken, the name 'arum' made its way to the British Isles very early on indeed, for it was even used in the native Anglo-Saxon medical literature. In contrast, the specific of its Latin name, *maculatum*, simply means 'spotted' or 'marked' and identifies the British plant from the many other species found around Europe and the tropics.

So much for speculative word play. Let's now leave the ivory tower of linguistic stepping-stones and enter the local tavern of folklore, for it is here, in the company of rustic straight-talking and pagan insight, that we meet the so-called common names that have given this plant one of its many claims to fame. They're a rough and ready lot it's true; in contrast to the scientific naming, which seeks to weed out ambiguity and create a certainty of specificity, the common names are a riot of individuality and regional creativity, of 'tell it like it is' and bawdy sexuality and, as with any crowd of well-natured reprobates, political satire. They sing out a verbal record of our own history as a nation and culture. The dictates of fashion, the rise of industries, the influence (and ridicule) of the church and our changing social norms are all reflected in the names Arum has inspired. Few if any plants have acted as such cultural reflectors, yet Arum seems continually to draw out and inspire our creativity for naming.

The hooded green cowl embracing the upright red poker has such obvious sexual symbolism that it would be remarkable if the plant had not garnered such a collection of salacious titles. A great many are gender-based twin names such as 'Lords and Ladies' or 'Stallions and Mares', mirroring the different parts of the plant's 'flower' and alluding to its strange seeming male-female nature.

Many are more overtly sexual in their nature, such as the well-known 'Cuckoo Pint', even if that is not so obvious to us today when the original meaning of the words has faded away. Pint, Pintle, and Point are all derived from the Anglo-Saxon word for penis and Cuckoo is from cucu, meaning lively (in a phallic sense). Clearly our ancestors pulled no punches when expressing what this plant reminded them of. A common element in many of Arum's names is the epithet 'Robin' and though we do not recognise its impudent rudeness today, once upon a not-to-distant-time-ago, oh but we did. It is from the French for cock; really meaning like a tap but used for both types of curved plumbing and it has even made its way into Arum inspired literature from Elizabethan times, as we shall see. The various canine-related names such as Dog's Dick and Dog's Tassel are equally direct in their phallic reference to this plant's virile posturing. Arum clearly does not inspire subtle side references to what people thought this plant looked like.

Our rural ancestors didn't just think about sex though. Some names such as Toad's Meat, Adder's Meat or Adder's Victuals, contain a warning, informing us of Arum's worthlessness as a food plant. Other names reflect its use in industry, such as Starchwort and Portland Starch. Buckrams echoes its use as a source of stiffening agent (an unintentionally ironic yet appropriate use for Arum). The latter name seems to derive from buckram, which is a coarse fabric sized with glue and used in bookbinding. It is also related to Bukhara, an ancient town on the Silk Road which supplied exotic silks and textiles including stiffened linen. Arum was seemingly sufficiently well known to be used as a suitable substitute or, at least, for its name to be used as a marketing ploy.

Kings and Queens is said to refer to the different colours of the plant reflecting the different colours worn by royalty in Elizabethan times. The names beginning with 'Parson ...' and 'Devil's ...' reflect a satirical view of local churchmen and the growing influence of the church's teachings. Victorian inhibitions bestowed disguised sexual references and portrayed the contemporary popular fascination with fairies.

Some names show a fanciful creativity, such as the lengthy Kitty-Come-Down-the-Lane-Jump-Up-and-Kiss-Me, while others show humour such as Cobbler's Thumb, which effectively conjures up the comparison of the red spadix with the glowing thumb of a poor cobbler freshly hit with his hammer whilst mending shoes.

That this profusion of names doesn't appear to be any different across mainland Europe is illustrated by this extract from *The Names of Herbes*, by W. Turner from 1538:

'Arum is called in greke aron, in english cuckopintell, Wake robin or Rampe, in duche Psaffen bynde, in frenche, Vidchaen, the poticarie calleth it Pes vituli, serpentaria minor, luph minus, groweth in euery hedge almost in englande aboute townes in the sprynge of the yere.'

For the purposes of this book, when discussing arums in general, the name is never capitalised. When discussing Arum as a personality, it most definitely is. Arum or Wild Arum is the name of the subject of this book.

Next, let's take a joyful slide down the roll-call of Arum's folk-names; all 165 of them...

Further Reading

Grigson, G. (1955). An Englishman's Flora. From sources unknown, Grigson lists some of the counties of origin (or at least use) of many of the common names.

Cristhwaite, H. (2007). A Fire Not Blown. A great source of information on the roots of words and names.

CHAPTER 2

A RIOT
OF
NAMES

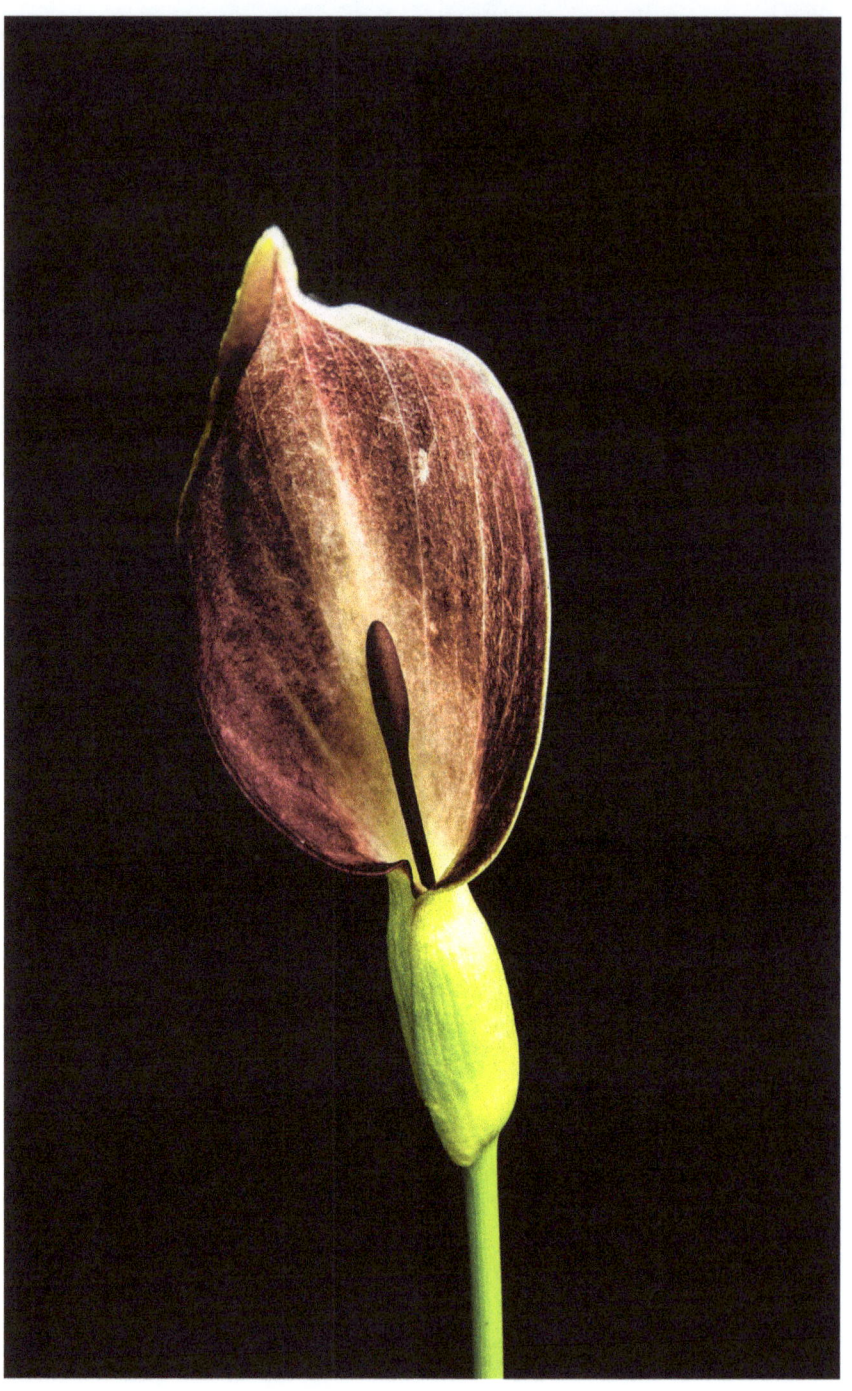

2

R eading like a register of rude or mocking phrases, the common English names for *Arum maculatum* show just how many identities this plant has in our psyche. Sources are mentioned for particularly modern sources (e.g. Mabey), where it is the only reference to this name I have been able to find or, in the case of names mentioned in Grigson's book[1], where a locality of origin or use is suggested. For the main part, names have been collected from the numerous historical documents and original herbal literature in which they occur. Only the English names are listed below but, for names of this plant in other languages, a very good source is Sue Eland (www.plantlives.com), where she provides many European common names for Arum (and many other plants), including the delightful Welsh *Pidyn y Gog*.

> Aaron.
> Aaron's Leek
> Adam and Eve (Grigson origin: Somerset, Leicestershire, Lincolnshire, Yorkshire)
> Adder's Food (Grigson origin: Somerset)
> Adder's Meat (Grigson says derived from the German for 'Snake Berries'. From Cornwall, Devon and Somerset)
> Adder's Root
> Adder's Tongue (Grigson origin: Cornwall, Somerset)
> Angels and Devils (Grigson origin: Somerset)
> Aron (Grigson origin: Scotland)
> Arum
> Arum Lily
> Arus
> Babe in the Cradle (Grigson origin: Somerset)
> Barba-aron
> Bloody Fingers (Grigson origin: Hampshire)
> Bloody Man's Finger (Grigson origin: Somerset, Worcestershire)
> Bobbin and Joan (Grigson origin: Cornwall/ Northamptonshire)
> Bobbing Jane

---

[1] *Grigson, G. (1955). An Englishman's Flora.*

Bobbins
Boys and Girls
Brown Dragons
Buckrams
Bullocks (Grigson origin: Somerset)
Bulls and Cows (Grigson origin: Somerset,
Northamptonshire, Lincolnshire, Lancashire, Yorkshire)
Bulls (Grigson origin: Dorset)
Bulls and Wheys [whey means a heifer or cow], (Grigson
origin: Yorkshire)
Calfsfoot
Calves Foot (Culpeper. Grigson origin: Somerset)
Canis Priapus
Clapper in the Middle of the Hole
Cobbler's Thumb
Cocky Baby (Grigson origin: Isle of Wight)
Cows and Calves (Grigson origin: Devon, Dorset,
Somerset, Wiltshire, Gloucestershire, Buckinghamshire,
Northamptonshire, Warwickshire, Worcestershire,
Shropshire, Nottinghamshire, Lincolnshire, Yorkshire, Lake
District)
Cows and Bulls
Cows and Kies (Grigson origin: Yorkshire)
Cow's Parsnip (Grigson origin: Somerset)
Cuckoo Cock (Grigson origin: Essex)
Cuckoo Flower (Grigson origin: Northamptonshire)
Cuckoo Pint [pronounced to rhyme with mint rather than
'pint' as in beer, this is from the Anglo-Saxon word 'pintle',
meaning penis] (Culpeper. Grigson origin: Sussex, East
Anglia, Northamptonshire, Leicestershire)
Cuckoo Point (Grigson origin: Yorkshire)
Cuckowe Pyntyll
Cypress Powder
Dead Man's Fingers (Grigson origin: Worcestershire)
Devils
Devils and Angels (Grigson origin: Dorset, Somerset)
Devil's Ladies and Gentlemen (Grigson origin: Denbigshire)
Devil's Men and Women (Grigson origin: Shropshire)
Dog Bobbins (Grigson origin: Northamptonshire)
Dog Cocks (Grigson origin: Wiltshire)
Dog's Dibble (Grigson origin:Devon)
Dog's Dick

Dog's Spear (Grigson origin: Somerset)
Dog's Tassel (Grigson origin: Somerset)
Dragon Root
English Passionflower
Fairies (Grigson origin: Somerset)
Fairy Lamps (Mabey: East Anglian Fens)
Figure of the Pestle
Fly Catcher (Grigson origin: Wiltshire)
Friar's Cowl
Frog's Meat (Grigson origin: Dorset)
Gaglee
Gentleman's Finger (Grigson origin: Wiltshire)
Gentlemen and Ladies (Grigson origin: Oxfordshire)
Gentlemen's and Ladies' Fingers (Grigson origin: Wiltshire)
Gethsemane
Goat's Ear (Ireland)
Greasy Dragon
Great Dragon (Grigson origin: Sussex)
Hobble-Gobbles (Grigson origin: Kent)
Hooded Aron
Hooded Cuckoe Pint
Jack in the Box (Grigson origin: Somerset, Buckinghamshire,
Northern Ireland)
Jack in the Green (Grigson origin: Somerset)
Jack in the Pulpit (Grigson origin: Cornwall, Somerset,
Lincolnshire)
Janus
Karup
Kings and Queens (Grigson origin: Somerset, Lincolnshire,
Durham)
Kitty-come-down-the-lane
Kitty-Come-Down-The-Lane-Jump-Up-And-Kiss-Me
(Grigson origin: Kent)
Knights and Ladies (Grigson origin: Somerset)
Ladies and Gentlemen (Grigson origin: Somerset, Wiltshire,
Kent, Northamptonshire, Shropshire)
Ladies' Lords (Grigson origin: Kent)
Lady's Finger (Grigson origin: Wiltshire, Gloucestershire,
Kent)
Lady's Keys (Grigson origin: Kent)
Lady's Slipper (Grigson origin: Wiltshire)
Lady's Smock (Grigson origin: Dorset, Somerset,

Hampshire)
Lamb-In-A-Pulpit (Grigson origin: Devon, Wiltshire)
Lambs Lakens [Lakens means toys] (Grigson origin:
Northamptonshire, Northumbria, North England)
Lilly (Grigson origin: Wiltshire)
Lilly Grass (Grigson origin: Sussex)
Long Purples (Grigson origin: Warwickshire)
Lords and Ladies (Grigson origin: General, from Cornwall
to Lakes and Yorkshire)
Lords' and Ladies' Fingers (Grigson origin: Warwickshire)
Man In The Pulpit (Grigson origin: Somerset)
Mandrake (Grigson origin: Yorkshire)
Men and Women (Grigson origin: Somerset)
Moll of the Woods (Grigson origin: Warwickshire)
Naked Boys
Naked ladies
Narrow Ear (Ireland)
Nightingale (Grigson origin: Essex)
Old Man's Pulpit (Grigson origin: Somerset)
Oxberry (Grigson origin: Worcestershire)
Parisian Cypress
Parson and Clerk (Grigson origin: Devon, Somerset)
Parson in his Pulpit (Grigson origin: Devon, Dorset,
Somerset, Cheshire, Yorkshire)
Parson in his Smock (Grigson origin: Lincolnshire)
Parson's Billycock (county unknown, though Grigson refers
to Shakespeare's King Lear)
Passion and Clerk (Grigson origin: Devon. Somerset)
Passion Flower
Pintle,
Poison Fingers (Grigson origin: Dorset)
Poison Root (Grigson origin:Wiltshire)
Pokers (Grigson origin: Somerset)
Portland Arrowroot
Portland Sago
Portland Starch Plant
Preacher in the Pulpit (Grigson origin: Somerset)
Prestes Hood
Priest in the Pulpit (Grigson origin: Somerset)
Preisties (Grigson origin: Lancashire)
Priestes Pyntill
Priesties (Grigson origin: Lancashire)

Priest's Hood
Priest's Pilly (Grigson origin: Westmorland')
Priest's Pintle (Grigson origin: Derbyshire, Lincolnshire, Durham, Cumbria)
Quaker
Ramp of Aron
Rampe
Ram's Horn (Grigson origin: Sussex)
Ramson (Grigson origin: Cumbria)
Red Hot Poker (Grigson origin: Somerset)
Robin and Joan
Sacerdotis Penis
School Master (Grigson origin: Sussex)
Serpentaria Minor
Shiners (Mabey: East Anglian Fens)
Silly Lovers (Grigson origin: Somerset)
Small Dragon (Grigson origin: Sussex)
Snake's Food (Grigson origin: Devon, Somerset)
Snake's Meat (Grigson origin: Devon)
Snake's Victuals (Grigson origin: Wiltshire, Gloucestershire)
Soldier in a Sentry Box (Mabey)
Soldiers (Grigson origin: Somerset)
Soldiers and Angels (Grigson origin: Devon)
Soldiers and Sailors (Grigson origin: Somerset)
Stallions (Grigson origin: Lincolnshire)
Stallions and Mares (Grigson origin: Lincolnshire, Yorkshire)
Standing Gusses (Grigson origin: Somerset)
Starch Flower
Starch Plant
Starch Root
Starchwort
Sucky Calves (Grigson origin: Somerset)
Sweethearts (Grigson origin: Somerset)
Tender Ear (Ireland)
Toad's Meat (Grigson origin: Cornwall)
Wake Robin (Grigson origin: Cornwall, Dorset, Sussex, Berkshire, Warwickshire, Worcestershire, Cheshire, Yorkshire, Scotland, Northern Ireland)
White and Red (Grigson origin: Dorset)
Wild Arum
Wild Lilly (Grigson origin: Devon)

Willy Lilly (Mabey)

Yaro

There are two curious features about this list. Of all the 165 names, just one; Oxberry, refers to the berries of the fruiting stage of the plant and only three make the obvious link between the shape of Arum's spathe and an animal's ear. The remaining 164 names all revolve around the distinctive flowering structure.

Now that we have been introduced, it's time to delve into the background of this fascinating plant because Arum was there, right at the beginning, when the very earliest books on herbal medicine were being created.

CHAPTER 3
THE ARCHETYPAL ARUM

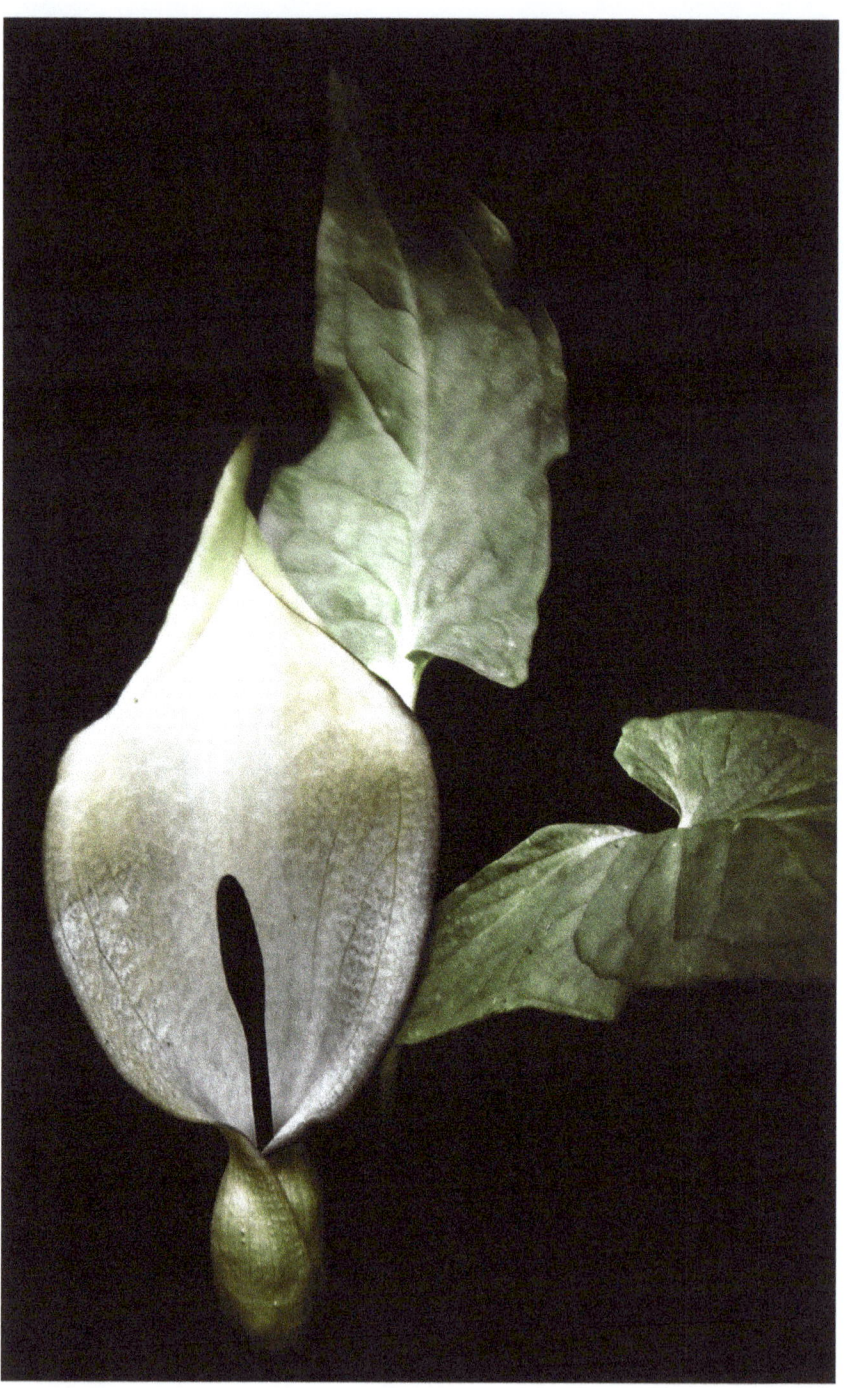

3

Befefore we step back in time some two thousand years, it's probably
wise to gird ourselves with some fore-knowledge of the times, so
that we are not led astray into the thicket of taxonomic confusion by our
modern mind-sets. In particular we need to be aware of how our ancestors
viewed and described the natural resources around them. Specifically, how
they dealt with a plant as tricksy as Arum. Let's begin with where we are
now.

Plant guides of today call the Common Arum which we see in the
British countryside *Arum maculatum*, describing it as one of the two UK
representatives (or species) of the genus Arum which is itself part of the
wider group of plants known as the Aroid or Araceae family. They may
go on to state that there are around 109 genera in the family and 25-28
species in the genus Arum, most of which are found in the 'Old World'
continents. Specialist guides to 'arums' may go on to describe how, in the
last few years, modern botanical taxonomy has redefined this group of
plants in increasingly specific and distinct detail, including entire removal
of many plants originally classed as Arum to other genera.

In contrast, the ancient herbals made no such detailed
distinctions. The term 'Arum' was used to denote any vaguely 'arum-
ish'-looking plant, just as someone today might use the term daisy or
dandelion to describe any daisy or dandelion-like flower, without knowing
the specific family or species. For most people, most of the time, this 'folk-
level knowledge' was and is good enough and generally tells us all we need
to know in order to recognise and use a plant. The ancient herbals take
the same view. As a result, the Arum we know in Britain (*Arum maculatum*)
is not specifically the one that appears in the herbals down the centuries.
At least, not all of the time. That's because the Arum of the herbals isn't
really any plant at all; it's a generic 'idea' of the 'Arum Plant': its type, even
its archetype let's say, rather than any specific species. It's all the arums
rolled into one.

This is perhaps not what we might expect from a guide book on
herbal medicine, at least not today. So why are the herbals so vague and
general, compared with our so-specific modern standards?

To answer this, we need to feel our way into the cultural and historical milieu in which the old herbal manuscripts were written. There we find that certain assumptions about the world which we take for granted, are quite different. While our modern-day society values original observation and new insight (and indeed actively punishes plagiarism), for much of our past the unquestioned truth was that the older the source the greater was its authority. In stark contrast to the iconoclasm of today, our ancestors displayed an attitude of subservience to tradition which meant that a great many assumptions went long unquestioned. The scientific approach to knowledge and enquiry we take for granted today took a long time to develop, with many setbacks and resurgences along the way. It was not until the 19th century that the sciences of taxonomy, botany, medicine and pharmacology really came of age as distinct and separate schools of enquiry and thought. Prior to this the lack of intellectual specialisation and the weight of historic authority meant that knowledge and presumptions about the world were rarely questioned in any tightly focused way.

As a result, the classical herbals are largely copies of copies, each in turn based on earlier herbals, back and back into time, with their origins in hand-written manuscripts written over two thousand years earlier. New information was added only sporadically and the texts of the original (and highly revered) authors rarely questioned. Herbals also tended to focus on those plants found around the Mediterranean region (including Egypt) where the original historical writers lived and travelled. Subsequent authors, even over a thousand years later, generally made no allowance for this because the concept that plants might vary according to geographical region was one which was very slow to catch on.[1] That there existed more than one Arum across Europe, for example, was a realisation long in the arriving.

There is also one more reason for there generally being just one generic or archetypal 'Arum' in the ancient herbals: the focus of these books was quite specific and different to that of plant guides today. Herbals were written to help people to treat illnesses. They were

---

[1]   *The concept had actually been pointed out already by Pliny in AD77, but it seems to have been overlooked, forgotten or ignored by subsequent writers and printers of herbals.*

not written as works of taxonomy or botanical classification and such distinctions would not have contributed to the usability of these books (even if they had been understood). If a group of similar-looking plants had similar acting medicinal properties, then for all practical purposes they were in effect the same plant, even if they sometimes looked just a little bit different. Because the regional variations, species and subspecies of Arum were perceived as all broadly similar when it came to their medicinal properties, it meant that in effect, they could be treated as one single plant – Arum. And so they were. And so will we, too.

In this celebratory exploration of Arum we are mostly talking about *Arum maculatum* and its close relatives but in reality a distinction is not tightly drawn and those taxonomic hairs are not so finely split. It's the spirit of Arum we are exploring here. Its inner personality that we are being introduced to. So now that we know what we mean when we talk of 'Arum' and we know what the ancient herbals mean when they write about 'Arum', let's take a journey into the past, back two thousand years, to the beginnings of what we would call a 'herbal' and the earliest written knowledge of medicinal plants.

Further Reading.

Arber, A. (1912). Herbals. Their Origin and Evolution.

Rohde, E. S. (1922). The Old English Herbals.

CHAPTER 4
THE ARUM OF THE MANUSCRIPTS

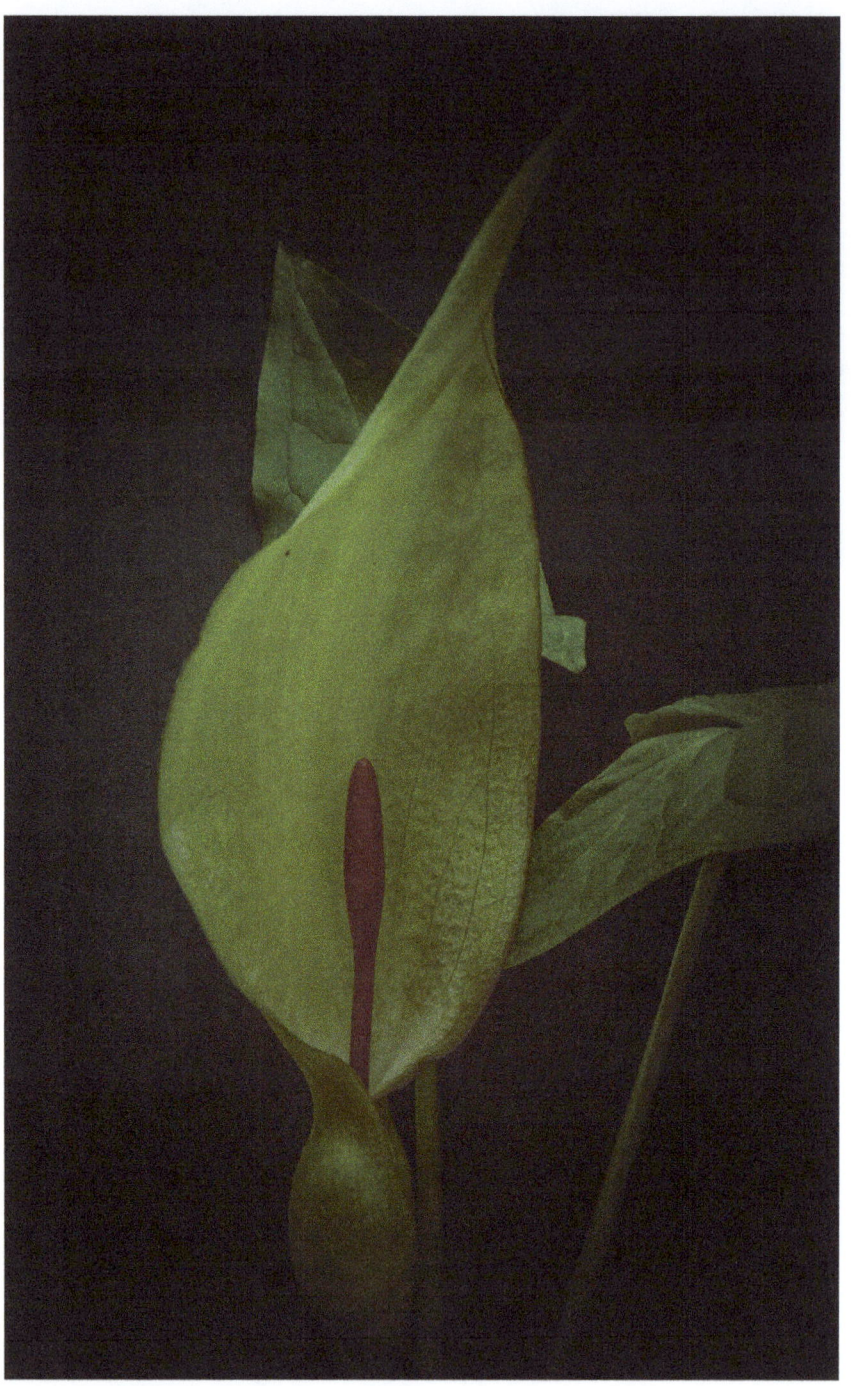

4

The earliest written records regarding the use of plants as medicine date back to the very dawn of civilisation itself. Sumerian clay tablets and Chinese manuscripts have been found from three thousand years ago bearing herbal prescriptions. The earliest complete medical book is the *Ebers Papyrus* from Egypt, which was created around 1550 BCE but contains herbal prescriptions from up to two thousand years earlier.

Arum first puts in an appearance in the early herbals and writings of Ancient Greece. It is from this time and culture that our knowledge of Arum's herbal properties has its origins, firmly embedded within the roots of our western culture. Three works by the founding figures of the great Mediterranean tradition of herbal medicine all mention Arum.

Theophrastus.

The first of these is by a student of Plato and Aristotle, named Theophrastus. Around 340 BCE he wrote *Enquiry into Plants*, one of the earliest written attempts to codify the natural world and a precursor to modern botanical classification. It was also the earliest written reference we have to Arum. In it, Theophrastus debates whether Arum really has roots in the same way that other plants do because instead of tapering to a point, the way that 'true roots do', they instead get wider the deeper into the ground they go. Hence, they cannot really be 'proper' roots. An interesting point for discussion perhaps, but he didn't really come up with any suggestions as to what else they might be, if not roots.

Theophrastus' work was not so much a herbal as a very early work of botanical enquiry about the nature of plants. It was an attempt to understand, mostly from a philosophical viewpoint, how plants 'worked' and how they could be classified. Sadly, as Arber says in her book *Herbals*, 'Aristolelian botany suffered from one serious handicap: an inadequate basis of actual fact'. She describes how Greek botany was created by philosophers who, being completely at home in the world of ideas, believed that any knowledge could be derived from thinking about 'general principles of the world' without any need for actual observation:

'… it was left for workers in the apparently less promising field of medicine' to do that.

Pliny's Encyclopaedia.

The second major publication of this time was the first ever encyclopaedia, written by Pliny. This really was an attempt to write out by hand a book of, literally, everything: a huge and momentous undertaking. Pliny's *Naturalis Historia* was published around AD 77 and is the only one of his works to have survived. It set the standard for all subsequent encyclopaedias as it contained an index (a very new concept); it referenced the original sources (a remarkable thing for its day and for some time to come, as we shall see), and ordered its contents in a way that made it accessible to anyone consulting it. It covered all that was known at the time, set out across almost 40 volumes:

Vol. 1: Preface and tables of contents, lists of authorities;

Vol. 2: Mathematical and physical description of the world;

Vols 3-6: Geography and ethnography;

Vol. 7: Anthropology and human physiology;

Vols 8-11: Zoology;

Vols 12-27: Botany, including agriculture, horticulture and pharmacology;

Vols 28-32: Pharmacology;

Vols 33-37: Mining and mineralogy, especially in its application to life and art, including gold, chasing in silver, statuary in bronze, painting, modelling, sculpture in marble, precious stones and gems.

Arum has a number of entries in Pliny's *Naturalis Historia* and, interestingly, he writes that already there is disagreement about whether Arum is one plant or many, on account of the different variations found around Europe, all quite similar but with noticeable differences. It is already known as '*Aron*', '*Dracunculus*' and '*Dracontium*', and the Arum which Pliny mostly discusses is not the British Arum but the Egyptian Arum: a plant now known as *Arum colocasia* or Taro. This is a notable example of plant observation (along with that from Egypt) which was subsequently forgotten in the later herbals.

Dioscorides.

Around the same time that Pliny produced the first volume of his great encyclopaedia, there occurred the third and probably most significant publication of this period. This was the creation of a herbal called *De Materia Medica*, written by a Greek soldier and army doctor known as Pedanios Dioscorides of Anazarba. This single work was to have more influence on herbal literature than any other for the next two thousand years.

Surprisingly, Dioscorides' work was not the first medical book produced by the ancient Greeks but it is the earliest one to have survived intact. The very earliest herbals that we know about are all now extinct, in that the manuscripts themselves have long since disappeared and they have become 'ghost works', their only remaining existence being fleeting references or tantalising excerpts found in the works of others. There was, for instance, Diocles of Carystos, a student of Aristotle known as the 'Second Hippocrates' by his contemporaries. He wrote a large number of books including the first known book on anatomy (according to Galen), a book on sexual health (possibly the earliest self-help book ever written) and numerous books on particular issues of health and specific diseases. He was also the very first Greek author to produce an actual herbal, written around 350 BCE.

Another 'ghost work' was *The Rhizotomicon*, produced in 120 BCE by a Greek called Crateus, who was a physician to Mithridates VI, king of Pontus from 120-63 BCE. About this herbal we know very little, though it was to prove highly influential. Its most notable feature was that it contained full colour illustrations of the plants it covered. This in itself was ground breaking, thoroughly unique and hundreds of years ahead of its time. Its innovative features didn't stop there. Along with the expected instructions on the medical conditions each plant treated, it also contained descriptions of the plants themselves along with details of how and where they grew. This was much more specific than anything which had gone before and, in many ways, of much that was to come after for a very long time. Of the text of this book, quotations by Dioscorides form our only remaining knowledge. We know of it primarily because many of the drawings and coloured illustrations in subsequent herbals over the next one thousand years are thought to be copies of Crateus' original creations.

While not then the first such work, it was in Dioscorides' great manuscript that the medical herbal found its foundation. Dioscorides had travelled and practised widely for many years in his career as a physician and soldier before setting down his experiences on parchment around 65 CE. Dioscorides incorporated material from a number of earlier authors such as Theophrastus, Crateus, Diocles and from another work now disappeared: the *Herbal of Sextus Niger*. That said, this was not just a compendium of earlier herbals. Dioscorides' five-volume work included the results of his own investigations, experience and observations and contains details of around 600 different plants, which is around 100 more than anyone else had previously described. He also presented a system of classification which was essentially pharmacological, grouping the plants together according to their medical properties. This was so far ahead of its time that, in subsequent copies, scribes ignored Dioscorides' ideas and reverted the manuscript to the traditional alphabetical order of listing the plants. So much for innovation.

The original manuscript of Dioscorides has long since disappeared. The earliest complete version to have survived is a manuscript from 512 CE known as the *Juliana Codex* or the *Codex Vindobonensis*. Few other books have such an illustrious history.

It was produced as a gift by the local townspeople of Constantinople for Juliana Anicius in response to her construction of a local church. Juliana was the daughter of Flavius Anicius Olybrius, who was briefly the Roman Emperor of the Western Empire in 472 CE.

The manuscript is around a thousand pages in length and magnificently illustrated, with almost 400 full-page colour paintings opposite the plant descriptions. Already, the plants have been arranged alphabetically, ignoring Dioscorides' original classification system. Extra material from other ancient authors has also been added, including a guide to over 40 Mediterranean birds - not what one would expect to find in a herbal and definitely not part of Dioscorides' original text. Many of the illustrations though are thought to be copies of those found in the now extinct *Rhizotomicon* of Crateus.

Following its creation and presentation to Juliana Anicius, the manuscript then disappears from history. We don't know who owned it or to what countries it travelled, yet a measure of how useful and valuable it was considered is that by the time it resurfaces, over a thousand years later in 1652, its parchment pages are teeming with handwritten notes and amendments in over 20 different languages including Arabic, Turkish, Hebrew and French. It had clearly passed through a wide variety of privileged hands. Remarkably, rather than sitting forgotten on the dusty shelves of a library, this single individual book had been in constant use for over a thousand years.

This use would seem to have taken its toll. At the time the book resurfaces into written history, it is the property of a physician in Constantinople who received it from his father; the personal physician of Süleyman the Magnificent (after the city came under Turkish rule in 1453). The manuscript was described as being in such a bad state that 'no one, if they saw it lying in the road, would even bother to pick it up'. These were the words of an ambassador of the Roman Emperor Ferdinand I, who wished to buy the manuscript but could not afford the 100 ducats being asked for it.

Such was the pull of this work though that by 1659, only 7 years later, the Emperor Maximilian II did buy it, to be held by the Austrian National Library in Vienna, where it has remained to this day.

Though *De Materia Medica* was translated into a great many mainland European languages throughout the centuries, it was not until 1652-5 that a John Goodyear produced an English version, writing below the original Greek with a line-by-line English translation. The book took him 3 years to write and filled over 4000 pages, each one handwritten, yet strangely, it did not see the light of day until 1933, when it was finally published after being rediscovered in an Oxford library. Incredibly, this was the sole translation into English until a completely modern translation was published in 2000.

Dioscorides' *De Materia Medica* was effectively the last word in herbal books for the next 15 centuries, dominating the contents of virtually all subsequent publications. It was the one source to which anyone aspiring to be, or working as, a herbalist would refer. In fact, despite that many of the medical recipes contained in it would not now be considered effective, it would be so slavishly copied, referenced and referred to that no real developments took place in the science of herbalism for the next 1500 years, because no one thought that anyone could do anything better. To question Dioscorides was unthinkable.

Not everyone could afford to have their own handwritten (or later on, printed) copy of Dioscorides' huge work and it fell to others to fill the gap in the market for more utilitarian 'workbook'-type herbals for the general populace. One such work was the incredibly successful *Herbarium of Apuleius Platonicus*.

This herbal is based primarily on work from Dioscorides, Pliny and Hippocrates and has 113 illustrated chapters, including one on Arum. The true identity of the original author (Apuleius Platonicus being only a pseudonym) has been lost over time, as has the origin of the book itself. The earliest copy existing dates from the 1400s and is based on a manuscript from the 9th century. The manuscript itself is thought to have been originally written as early as the 4th century in Greece.

The *Herbarium of Apuleius Platonicus* is possibly the most popular herbal manuscript in existence, continually 'in print' through being repeatedly copied by hand for over one thousand years. With the arrival of the printing press, production of this book went into overdrive. Over 60 manuscripts from the medieval period alone still survive. It has the further distinction of being the first printed herbal with illustrations and possibly the first of the Mediterranean herbal literature to be translated into English. Yet it was still primarily based on Dioscorides' work and was effectively a 'popular' version of his 'master' manuscript.

In looking at the history of herbals in the West, we are traditionally steered through that fountain of knowledge and writing which springs from Ancient Greece and becomes the established and mainstream tradition across the Middle East, Europe and the Near East from 1500 BCE onwards. Interwoven with this, however, is another stream of herbal knowledge, largely unrelated to this Greek and Mediterranean tradition. It comes from this strange and faraway northern land now known as Britain. Let's take a brief detour to see what was happening up in Britain around the 9th century.

Further Reading.

Pliny's Encyclopaedia: bit.ly/Yh9Aud

Dioscorides. (2000). Osbaldeston, T. A. IBIDIS Press. A modern translation.

CHAPTER 5

THE ARUM OF THE ANGLO-SAXON MAGICAL HEALERS

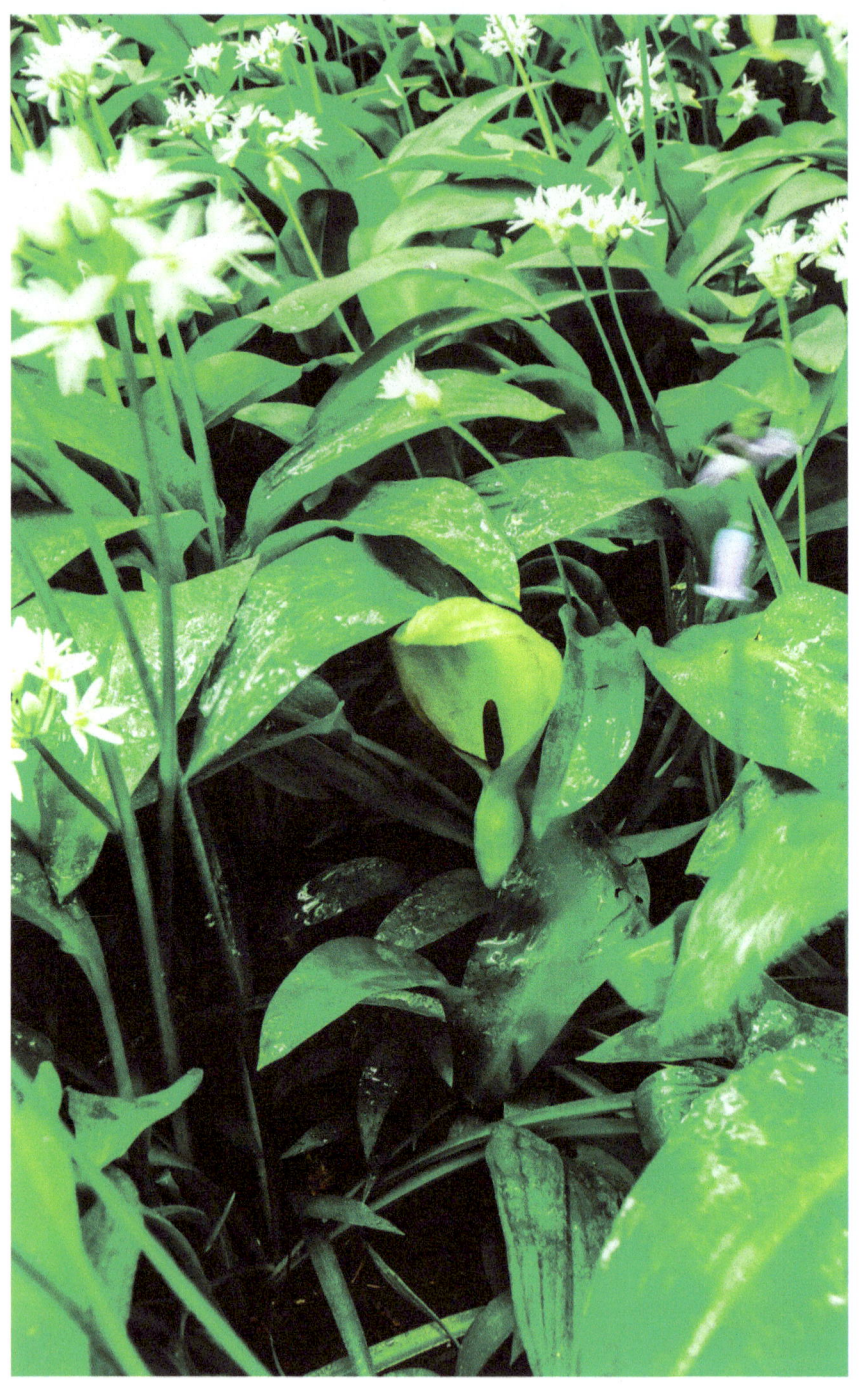

5

O f our Anglo-Saxon herbal knowledge and practices, only a
fraction has survived to this day. Most of what was known was
never written down and much of what was has been destroyed, either
deliberately by the Normans or by the simple passage of time. What
remains shows a world of great herbal knowledge intimately intertwined
with ancient rituals, deep superstitions and a firm belief in the elven folk.
It is a tantalising slice of a world now long retreated into the wilds of
history.

Two main works of medical practice stem from the British Isles
and this distant period: the Anglo-Saxon *Leechbook* and the manuscripts of
the Welsh Physicians of Myddfai. Both manuscripts are a written record of
an oral tradition stretching back into distant and pre-literate history.

Bald's Leechbook.

Of the two, the Anglo-Saxon *Leechbook of Bald* represents the oldest
surviving British medical textbook we have and is a compilation of medical
practices current at the time of Alfred the Great. It is called 'Bald's'
leechbook because Bald was the person who ordered it to be compiled,
by a scribe named 'Cild'. Contrary to today's usage, the word 'leech' in
the title does not refer to the bloodsucking variety, but is the modern
rendition of the Old English word *laece*, the word for a healer. This in
turn is derived from the root word *laekjaz*, meaning enchanter or 'one
who speaks magic words'.[1] *Laece* is also related to another Old English
word, *lac*, which relates to the practice of sacrifice and the offering of a
gift or wealth. Here we come upon a very ancient association between
magic, healing and the making of sacrifices, either to the Gods or to the
'leechmen' in return for health. With the Old English word *laece-feoh*, or
'leech fee', it illustrates the long-standing power of healers and doctors to
extract wealth from patients – an ability retained to this day.

Bald's *Leechbook* is, of all of the surviving Anglo-Saxon medical
literature, the most untouched by the otherwise all-encompassing

_____

[1]   *Pollington (2000)*

influence of the Mediterranean traditions. Split into three different books, the first two cover the treatment of external and internal problems, respectively, with both books arranging the remedies in a head-to-foot fashion. The third book is a collection of magical charms, rituals and folklore as medical remedies, which stem from a primarily English bedrock and show little of the Mediterranean influence so prevalent in other herbals. The herbal charms, superstitions and rituals recorded do not date from the 9th century but instead are an echo of a much older, pre-Christian time, dating back to the days of Beowulf. It is a curious mixture of Christian beliefs and pagan practices. Many illnesses were held to be the result of elf-shot, so much so that the longest chapter in the third book is entirely devoted to such elf-caused maladies. These were not the tiny gossamer fairies of Victorian times, but dark and powerful beings of nature who, more often than not, were unfriendly towards mankind. It was a 'time when grown men believed in elves and goblins as naturally as they believed in trees'.[1]

One of the notable features of the book is that it makes references to earlier 'leechmen', naming some of them and the remedies which they taught. Such references indicate that the records we have in Bald's *Leechbook* are based on an even older tradition of healing which seemed to have already organised itself into a kind of professional body with its own 'authorities': an insight into Anglo-Saxon society for which we have no other evidence. Even more remarkable is that the *Leechbook* is written in vernacular Anglo-Saxon, not Latin, demonstrating that there was already a section of society that was literate yet separate from the Roman–Latin tradition of mainland Europe. That their knowledge was extensive is evidenced by the fact that even the very limited Anglo-Saxon literature which has survived mentions around 500 different plants, a number exceeded only by Dioscorides. Even famous European herbals such as the *Herbarium of Apuleius* described only 185 different plants and this was one of the most popular herbals in Europe. Sadly, most of the Anglo-Saxon literature, medical practice and culture was to be destroyed only a short time after the *Leechbook* was written; not by the practice of Christianity, though that was playing its part, but by the coming of the Normans from France. The *Leechbook* is one of the few remaining windows we have into a society and world which no longer exists. After being held for a time at the

---

[1]   Rohde (1922)

Abbey in Glastonbury, this great compendium of the beliefs and practices of our ancestors from before the coming of Christianity now survives as just a single manuscript held in the British Library.

The Physicians of Myddfai.

The other healing tradition of which we have written record is that of a magical healing family from Wales. The Physicians of Myddfai were a family of healers and physicians from the parish of Myddfai in Carmarthenshire in Wales. They were said to have descended from three sons whose mother was one of the fairy folk to whom their father was married until she returned to her lake. They practised their fairy-derived skills from at least the 13th century as a family of healers. The last member of this family died only in the late 1700s, and their gravestones can still be seen in the parish churchyard today. The last descendent was John Jones, a respected surgeon, who died aged only 44 in 1789. The manuscripts were written down only towards the end of their line and represent a distinct and separate and equally ancient, healing tradition of these lands. They contain, like the Leechbook, a mixture of herbal cures and deep superstitions as well as strange rituals of cure.

Arum is mentioned in both the *Leechbook* and the recipes of The Physicians of Myddfai. Both record its use for dissolving growths and blockages in the body and the Leechbook entry in particular provides us with the earliest mention of Arum in English literature, where it is known by the name of Arod.

Though in many ways a minor side note in the great history of herbals, the British tradition is important for providing us with a small number of manuscripts that give us just a glimpse of our native Anglo-Saxon medicinal practices and traditions; before Christianity, the Normans and the Mediterranean tradition of herbal knowledge arrived at our shores and swept it all away.

After this brief detour into British herbal history, we must now return to the continent, for a startling new invention has appeared which is about to change everything and with revolutionary speed: the printing press.

Further Reading.

Pollington, S. (2000). Leechcraft, Early English Charms, Plantlore and Healing.

Cockayne, T. (1864). Leechdoms, wortcunning and starcraft of early England. Available from the Internet Archive: bit.ly/ZcXUdR

The Physicians of Myddfai. (1861). Internet Archive: bit.ly/14gp2dF

CHAPTER 6
THE AGE OF THE HERBALS

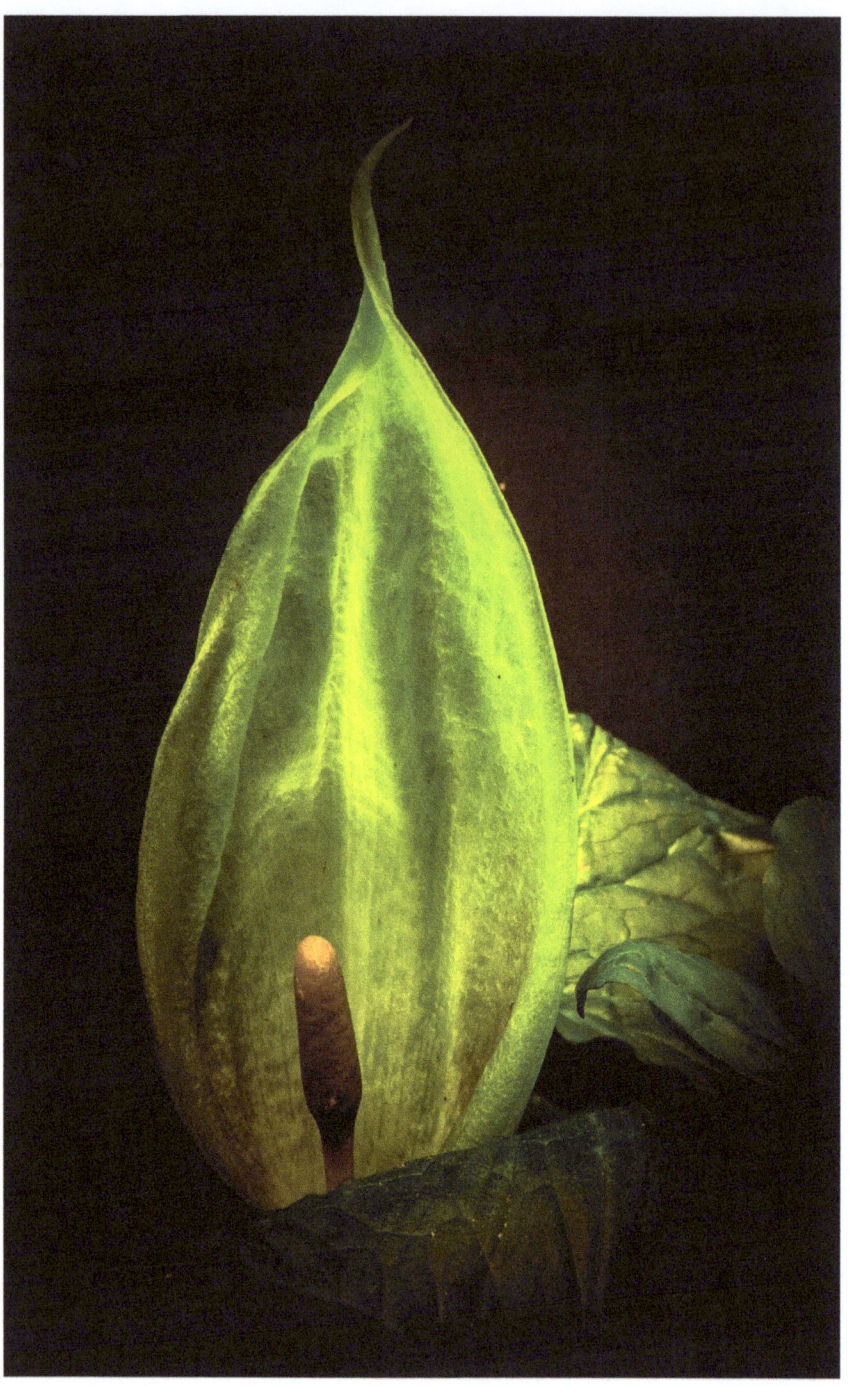

6

A natural way-point in our exploratory stroll through the history of Arum and herbals is the invention of the printing press in the mid-1400s. From here we can see its appearance for the historical singularity that it was. Prior to its invention, herbal manuscripts (in common with every other kind of book) were mainly reproduced only by being laboriously and meticulously copied by hand. From earlier than 400 years BCE to the mid 1400s, important herbal books were reproduced by being copied, written, drawn and bound by an individual's hand. They were rare and valuable items and each one at least a little bit unique.

The printing press changed all of this. It was a new industry, creating a new world and ushering in new possibilities. In every field of endeavour the relative ease of printing meant that existing manuscripts along with bright new works could suddenly be mass produced, expanding the accessibility to knowledge to more and more people. The new masters of the art - the printmakers - were quick to realise its potential.

The First Printed Herbals.

The first herbals produced with this new technology were often simply printed reproductions of works which had already been in existence in manuscript form for hundreds of years. They are known collectively as the 'Incunabula Herbals', incunabula meaning 'wrapped in swaddling clothes', and so rather lovingly referring to works produced in the first 50 years of the printing press: the newborn books.

The sudden ability to reproduce these ancient manuscripts, effectively en masse, was to turn the status quo of the previous two thousand years on its head. For the first time people were able to produce a copy of Dioscorides' great work effectively on demand, in as many copies as desired, in relatively no time at all (compared with how long it would take a scribe to copy out the entire manuscript afresh) and without having to pay a scribe for his time either. It must have seemed incredible, dangerous and exciting all at once. These were heady times for the printers and they lost no time in letting it go to their heads.

Quick on the uptake and hot off the presses was the Roman printer and diplomat Johannes Philippus de Lignamine, who issued the very first printed herbal in around 1480; a copy of none other than the herbarium of *Apuleius Platonicus*, the perennially popular anonymous herbal from the 5th century. This was closely followed by a series of books from Germany, all issued by the same printer, Peter Schoffer, between 1484 and 1485. The first of these was the *Latin Herbarius*, which seems to be based on a previously unknown manuscript. This was followed by the *German Herbarius*, the illustrations of which were of such quality that it became one of the books most copied by other authors. (In fact, its illustrations were not bettered for almost another one hundred years.) The third, the *Ortus Sanitatis*, was based on the previous *German Herbarius* but also included chapters on birds, animals, stones and fishes. It too was illustrated, with the images often attempting to convey a moral tale wrapped around the properties of the plants or semi-mythical animals being described. Peter Schoffer clearly knew a profitable business idea when he saw one.

New writers were also quick to get in on the act, though all was not rosy for such early authors. Just as early manuscripts were copied by hand from manuscript to manuscript, the printing press only amplified this time-honoured means of reproduction. Within months of a popular work being printed, copies were produced by other print houses and they seldom gave any credit for their original sources. Many writers and publishers would produce so-called 'new' works which were predominantly reworkings of existing publications or even just translations of works in other languages, producing a rolling tumbleweed of copying, mis-translations, half-truths, garbled medicinal information and, with few exceptions, illustrations of a steadily deteriorating quality. Plagiarism was the name of the day and the copying of others' work clearly has a very long precedent indeed.

To give an example of this, consider a very popular and high-quality herbal published in London in 1525 by Richard Bankes. Seemingly an original work – or at least based on no known earlier manuscript – it ends with the words 'Imprynted by me Rycharde Banckes, dwellynge in London, a lytel fro ye Stockes in ye Pultry'. It was the first herbal published in England and contained useful botanical information on the plants and their medicinal properties and often gave a greater amount of factual information than many other existing and more famous herbals. It also must rank as one of the most pirated books of the 16th century. Over the next 25 years an impressive number of different editions of Richard Bankes' work were produced. Unfortunately not one of them was by Richard Bankes. These copies were issued by other print houses and, to disguise their provenance, they were given different titles and ascribed to different authors. Sometimes they even claimed to be works by classical writers from Ancient Greece or the personal practice notes of eminent doctors. Once inside the covers however, they were simply copies of poor Richard Bankes' herbal.

Printing Comes of Age.

The sheer quantity of books produced even during the early years of the printing press is clear from the number still in existence. As Minta Collins writes, 'From the 15th Century alone, 193 different herbals have survived ... More than the number of herbals produced during the previous 900 years'. This alone illustrates the impact that the printing press was already having. Thus was born the great revolution that culminated in the so-called 'Great Age of Herbals' during the Elizabethan era of the 16th and 17th centuries. This was a time of great expansion and experiment across many different cultural fields and a time when many of the most famous and landmark herbals by writers such as Fuchs, Turner, Lyte, Gerard and Culpepper were produced, providing, eventually, the foundations for the modern scientific disciplines of medicine, botany and pharmacology.

Back in the 1500s however, such heady intellectual disciplines were still in the process of being forged amongst the pages of herbals by writers such as Leonard Fuchs. In 1542 he produced what was clearly a labour of love in the form of the most beautifully and accurately illustrated herbal ever produced, his *Historia Stirpum*. What made his herbal so outstanding and ground breaking was that each picture in his herbal was from an original woodcut based on an actual living plant. While today we would not expect illustrations to be based on anything else, in Fuchs' time, this represented an incredibly unique effort and a marked break with tradition. This was indeed his intention. In not allowing his craftsmen to 'indulge their whims', he desired to produce a book which would genuinely help people to identify the plants described. In many ways, his success in this is notable in that this was primarily a picture guide book to plants. The descriptions of the herbs and their uses was still based largely on classical sources, annotated where possible with Fuchs' own knowledge, though in practice this meant plants which grew near to where he lived in Germany.

In 1551 another physician, William Turner, published the first of a two-volume herbal: *A Newie Herball*. Though it copied most of its illustrations from Fuchs' book, Turner did supplement the text with the results of his own observations from his extensive travels and, very importantly, he wrote in the vernacular English, not in Latin. This was an unusual step for the time and took current herbal knowledge out of the hands of the professional apothecaries and physicians and made it accessible to the general populace of the land – a controversial and rebellious act that was typical of William Turner. When not busy occupying himself annoying the church authorities, he was well respected and his herbal writings led to him being known as the 'father of modern botany'.

In 1554 a very important work was produced by Rembert Dodoens, the *Cruydeboeck*. This was an illustrated herbal that again mostly copied its pictures from Fuchs' work, which had not long been published. In this case the copying was quite literal, as Dodoens used the very same woodblocks to reproduce the illustrations exactly. However, the text relates often to plants found around the Flemish area where Dodoens resided. He later produced a revised version with completely new illustrations (though perhaps only because Fuchs' woodblocks had by this time come to England where they were being used to make an English copy of Dodoens' work).

Dodoens' herbal was also important because it formed the basis for two herbals by English authors who both became very well known after writing their 'own' herbals, which were essentially copies of Dodoens' work.

In 1578 Henry Lyte published *A Newie Herball*, the title of which was doubly ironic for not only was Lyte's herbal itself not new but a translation of Dodoens' work into English, but even the title itself was the same as Turner's earlier work. Lyte somewhat escapes complete plagiarism if we indulge the full title of his book, which was '*A Niewe Herball or Historie of Plantes: wherin is contayned the whole discourse and perfect description of all sortes of Herbes and Plantes: their divers and sundry kindes : their straunge Figures, Fashions, and Shapes: their Names, Natures, Operations, and Vertues: and that not onely of those which are here growyng in this our Countrie of Englande, but of all others also of forrayne Realmes, commonly used in Physicke. First set foorth in the Doutche or Almaigne tongue, by that learned D. Rembert Dodoens, Physition to the Emperour: And nowe first translated out of French into English, by Henry Lyte Esquyer*'. Snappy. And definitely new.

Lyte did a good job with his translation though, making corrections and including information from other authors, including material sent by Dodoens himself. Nevertheless, Dodoens must have been just a bit regretful that Lyte now had Fuchs' woodblocks for making prints.

The next great copy of Dodoens' work was Gerard's herbal.

Gerard's Herbal.

Gerard's work was essentially a plagiarisation of a plagiarisation, being a reworking of an unfinished translation of Dodoens' revised work. Gerard finished the translation (after the previous translator had died), rearranged the contents and then put his own name on the cover. This in itself wouldn't have been so different from what everyone else was doing except that not only did Gerard claim that the previous translator's work had been lost, implying that this was all Gerard's work but, due to his own lack of plant knowledge, he mixed up the illustrations and the descriptions. This resulted in many descriptions of the herbs being illustrated with pictures of unrelated and random plants. Gerard did, however, add a considerable number of new plants to Dodoens' original work. Unfortunately, these were usually the rare exotics that Gerard grew in his own garden in central London and which no one in England would ever come across elsewhere.

To completely remove any aspect of reliability he or his herbal might have, Gerard also included myths and legends about plants presented as actual fact. One example of this is when he describes in his herbal the mythical 'barnacle tree'. This concerned those trees which grew large barnacles on their branches out of which, when they opened, hatched barnacle geese. Most other authors of the time related this as a 'commonly told tale or belief', but not Gerard. He wrote in his herbal that he had actually seen this with his own eyes. In Lancashire, he relates, he had seen the birds hatching out of these barnacles growing on old ships. Strange then that over 200 years previously, the great Albertus Magnus had already disproved this story by observing geese hatching from, of all unlikely things, eggs.

Incredibly, despite this insalubrious foundation, *Gerard's Herbal* gained a considerable following and in 1633 it was reissued, with better pictures which were now placed alongside the plants they were illustrating and with updated textual descriptions of the herbs and their uses. Carried out by a respected and competent botanist and apothecary, Thomas Johnson, this elevated Gerard's herbal to the comprehensive and genuinely useful herbal encyclopaedia that it should have been originally.

Culpepper's Herbal.

The last of the great herbals was Culpepper's *Complete Herbal*. Although his work is in some ways not terribly respected because it is immersed in the astrological symbolism of the herbs, Nicholas Culpepper himself was a remarkable person. He broke with the medical establishment almost immediately in his career by setting up his apothecary surgery just outside London, so that he was not under the city's authorities and could not easily by controlled by them. He was vocal in his beliefs that medicine should be affordable and accessible by all and often gave his services for free (having married a wealthy bride), sometimes seeing over 40 patients a day. He criticised fellow physicians for selling expensive remedies when herbs were available free from nature. He wrote his *Complete Herbal* in vernacular English rather than in Latin, describing native and not unobtainable exotic herbs (both features contrasting markedly with Gerard's herbal) and deliberately sold it very cheaply, resulting in it being, at the time, the single most successful non-religious English text ever written

Nicolas Culpepper led an extraordinarily eventful life, marked by great tragedy, huge privilege and fortune, a deep desire to help his fellow man and a marked defiance of authoritarian restrictions. One of his trainers in herbal knowledge was Thomas Johnson; the man who had so successfully rewritten Gerard's herbal. Culpepper was an incredibly successful and popular herbalist, yet he found no favour with the medical establishment whom he constantly attacked. When the Civil War came he fought on the side of the Parliamentarians (naturally enough, given his anti-royalist beliefs), receiving a wound from which he never fully recovered, though his prodigious workload and literary output probably did not help. He died in 1654 aged only 38, having changed the practice of herbal medicine forever. A comprehensive account of his life can be found at www.skyscript.co.uk/culpeper.html

From the 1700s onwards, what we would recognise as scientific enquiry began to emerge and have an impact on what was being written concerning the medicinal properties of plants. The herbals of old began to disappear and botany and medicine began their own journeys as new and separate disciplines. Let's now take a look at this incredible plant these herbals are describing and the properties they ascribe to it.

Further Reading.

As well as the books mentioned after Chapter 3:

Culpepper: www.skyscript.co.uk/culpeper.html

Collins, M. (2000). Medieval Herbs, the Illustrative Tradition.

The History of Modern Herbalism: bit.ly/XLIfUL

CHAPTER 7
ARUM THE TRICKSTER HEALER

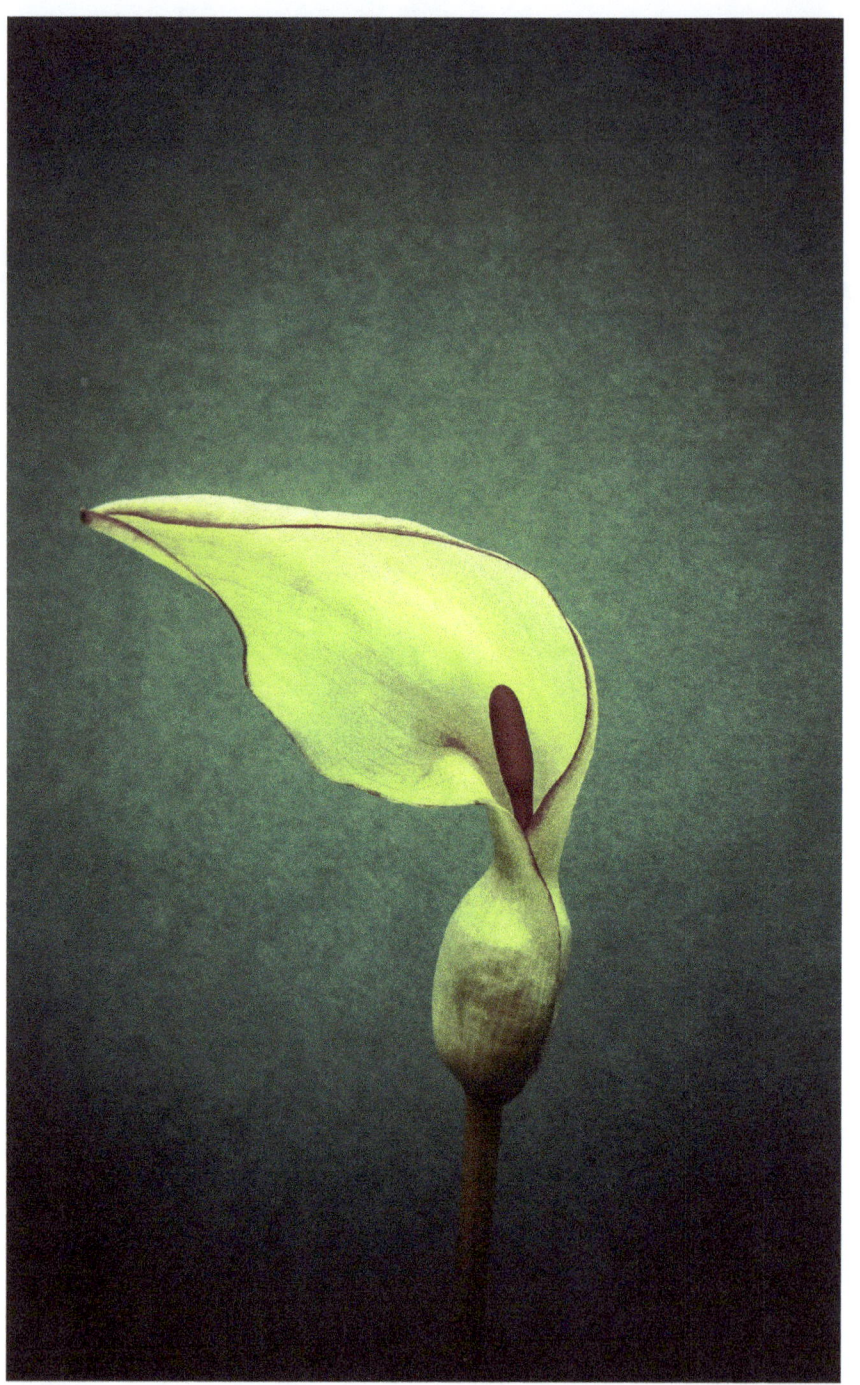

7

O pen the pages of any ancient herbal and there you will find Arum, its unique appearance marking its place in the parchment with its virtues described and laid out before us. But how different today. Leaf through any recent book on herbs and you will look in vain for any trace or mention of Arum. It has gone. It has become absent from our modern-day world of herbal medicine with its tame plants and safe remedies; banished from the books and erased from history in a seemingly determined effort to hide this plant's powers from common knowledge. Perhaps this is prudent; after all, this is not a plant for children to play with.

The story of Arum's disappearance from medical herbalism is that of the archetypal fall from grace of one whose powers are so strong they become their very downfall. Arum presents the paradox of the trickster: useful to humanity but also dangerous; powerful herbal medicine to those who know how to invoke it correctly, a deadly and painful poison to those who don't. To dare to seek Arum's herbal gifts transports us into the realm of the hero quest, the prize available only to those able to dodge the traps and dangers Arum presents to us. Today this quest has been judged too dangerous and has been largely forgotten. The tale of Arum is a tale of a reversal in fortunes from well-known centre stage to forgotten periphery and of a transformation from powerful benefactor to dangerous villain. An archetypal tale indeed. This shift in how Arum is perceived is all the more remarkable given its inclusion in the very earliest of herbal manuscripts and its continued presence in such works for the ensuing two thousand years. Yet when it came, its slide into medical obscurity was both rapid and complete. Today it is regarded as a plant of such poisonous qualities that it is not even to be touched, let alone taken internally – a level of aversion we project onto very few plants indeed. How did it come to this?

Arum's medicinal properties can be summed up in two words: caustic and astringent. Arum is a powerfully purgative plant with a markedly acidic effect on the body. It warms and heats, dissolves blockages, clears growths and obstructions and removes unwanted and excess matter from our bodies. Consider it a botanical bleach and you will have a fairly accurate picture of the coruscating affect it has on our bodies. In view of this, it is unsurprising that although it has a long history of medical use, Arum has always presented something of a dilemma to medicinal practitioners, being at the same time genuinely useful as well as genuinely dangerous. Its powerful qualities are reflected in the main uses to which it has been put: as an abortifant, as a purgative and as a burner away of growths, tumours and stagnations. This isn't a plant for minor ailments; it's a plant to call on when you absolutely positively have to clear away every last unwanted growth, blockage and stagnation occurring in the body.

To Induce a Birth, To End a Birth.

From most points of view, pregnancy is one of the most important 'growths' the body can produce. For obvious reasons, herbs to 'move stagnations and growths' are avoided by herbalists when treating anyone who is pregnant. Arum is very effective at 'getting things moving' and it is clear from even the very earliest of written records that Arum was well known as an effective abortifant. Dioscorides provides a very straightforward recipe with which to bring to an end an unwanted pregnancy. Simply mix 30 seeds into a posca, which is a mix of sour wine or vinegar along with a variety of assorted medicinal herbs, and drink. Whilst this sounds like a fairly unpleasant potion to our modern tastes, at the time of Dioscorides a posca was a popular draught, particularly as a safer alternative to drinking water. The acidity of the wine or vinegar helped to kill any bacteria and the herbs helped to cover up the, no doubt unpleasant, taste. How many times a day this should be taken Dioscorides does not say but Arum being Arum, it is likely that just one dose will bring about the desired result.

In fact, Dioscorides considered Arum to be so powerful in this regard he cautioned that even 'the smell of the root or of the herb, is destructive of late conceptions' and that 'the smell thereof after the withering of the flowers, is destructive of embrya newly conceived'. While Dioscorides' warning against even smelling the flowers seems overcautious, recent research has shown that he was actually being very literally prescriptive.

Arum is well known for producing a rich mixture of volatile chemicals which smell akin to rotting meat or dung. One of the main constituents of this mix is a group of chemicals known as dimethyl oligosulphides, renowned amongst chemists for their foul smell due to the sulphur atoms they contain. It is these sulphides which are the likely cause of Arum's abortifant properties. A study in 1998 showed that women exposed to inhaled sulphide chemicals have a greater propensity to miscarry compared with those not exposed. So it might well be the case that even the smell of Arum flowers could potentially cause a miscarriage in those who are particularly sensitive or at risk of miscarrying.

The one unexplained aspect of Dioscorides' comment is his warning against the smell left 'after the flowers have withered'. On the face of it, this seems an unlikely danger. Arums produce their volatile compounds to attract flies as an aid to pollination. Once this is achieved the spadix (or flower, as Dioscorides would have called this) withers and the plant stops producing any scent. The withered plant is also not one that would attract a person to smell it, being rather unattractive in its appearance and looking nothing like a flower. However, given that Dioscorides' strange warning has been so closely confirmed by modern research indicates that we should not be quick to dismiss this ancient master of herbal lore. Perhaps the withered Arum 'flower' holds yet some potency we still don't recognise?

What brings about abortion also provides an effective inducer and so, making use of the same properties for related purposes, Arum has also been widely recommended as an aid in birth delivery, both of people and (other) animals: to expel the afterbirth and also to bring on stagnated menses. Such a collection of related uses indicates the strength of the purgative and astringent qualities of Arum when taken internally.

The Purger and The Cleaner.

It is as a remover of unwanted growths where Arum has found its most enduring medical purpose. Its purgative and caustic nature lends itself perfectly to the dissolving of lumps and bumps both in and on the body and this type of application forms the main usage of Arum down the centuries, with remedies and recipes recommending Arum for these kinds of ills continually appearing until as recently as the last few hundred years, long after most of its other medicinal uses had been abandoned.

Examples abound in the ancient herbals. Hippocrates prescribed the leaves to be applied to the skin for the ridding of abscesses. This was still being echoed as late as 1947, in an account of a child being cured of an enormous tumour on its arm. With amputation seeming the only remaining option, the family consulted a local herbal healer who applied a daily compress of fresh Arum leaves. Though no timescale is reported, the tumour was fully cured by this treatment. Dioscorides recommends that Arum be taken with honey and White Bryony for internal ulcers. Suppositories were also made from it for the treatment of fistulas, and the juice of the leaves has long been applied to the eyes to clear them. Even in the Anglo-Saxon Leechbook and the writings of the Physicians of Myddfai the astringent properties of Arum were known. The Leechbook gives the following recipe:

'If a strong potion lodge in a man and will not come away, taken the netlierward part of celandine and leaves of libcorn or arod (Arum), boil in ale, add butter and sale, give to drink a cup full of it warm.'

The *Physicians of Myddfai*, writing twelve hundred years later, recommended using Arum to treat a cancer:

'Take the root of the dragons (Arum), cut them small, dry and make into a powder, take nine penny weights of this powder, boil in wine well and give it to the patient to drink, warm, for three days fasting and it will cure him; and I warrant him he will not have it again.'

Dioscorides prescribed Arum for the treatment of wounds and surface ulcers on the skin, including the treatment of leprosy:

'Beaten small with white bryony and honey it cleans malignant and spreading ulcers. It cleans away leprosy when rubbed on with vinegar ... The leaves beaten small are profitably applied to the newly wounded.'

Interestingly, the saponins in Arum have now been shown to be highly effective antibiotics, being notably effective against two types of bacteria which cause skin infections: *Staphylococcus epidermis* and *S. aureus*. Dioscorides and the ancient herbal lore prove themselves once again.

Arum's caustic and purgative effect lends itself to use as a general detoxifier, particularly given its traditional recommendations for treating such conditions as gout, gallstones, rheumatism and urinary infections, along with being an expectorant for bringing up mucus from the lungs. Once again, modern study has discovered that Arum contains a group of chemicals called cucurbitacins. These have a wide range of biological functions but are known to be hepatoprotective, meaning they support the liver and its functions as well as assisting the circulation system in general.

The Serpent Plant.

Arum also has a long association with serpents and dragons, so it has traditionally been recommended as a remedy for snakebite, with persons so afflicted to take the plant in red wine. Even better, to avoid being bitten in the first place, was to rub one's body with the leaves of the plant mixed in oil. The reptiles would then flee at one's approach. Burning it would also keep them from one's room.

An Arum Medicine Cupboard.

The minute and detailed extent to which our ancestors made use of the plants around them is illustrated by Dioscorides' reference to the use specifically of the juice of the seeds of Arum which, when pressed out and mixed with oil, will ease ear pain and destroy polyps in the nose. The *Grete Herball* of 1526 recommends Arum for haemorrhoids, advising the sufferer to bathe in water in which the plant is steeped or, alternatively, to sit on a hot cloth wrapped around the herb. In 1542 Fuchs summarised well the plant's properties for users of his herbal:

'The aron has the property of dispersing, reducing, and cleansing; therefore it heals swellings of the ears, piles, strumas and hard tumors, removes deformities of the face and skin'.

Turner in 1551 further bore this out, stating that the root 'purges all the inward parts, breaking through all humours, it purges and scoures mightely all things that need scouring, including freckles'. He also repeats that the fresh leaves are effective in the closing of wounds.

Herbals later than the 14th century, including the famous herbals of the Elizabethan era such as Culpepper, Gerard, Coles, Lyte and Turner, still predominantly regurgitated the work of Dioscorides in one form or another, but they did sometimes add one or two innovations for Arum which seemed to be of their own invention. For instance, a number of authors mention in passing that cheese can be preserved by wrapping it in Arum leaves. Lyte, in his herbal of 1578, states that the roots boiled with honey and then licked are good for those that 'cannot fetch their breath'. This is mentioned again by Culpepper, advising those who are 'short winded' to take the roots as a 'licking electuary', an electuary being a medicinal paste made from the plant along with a sweetening substance for improved palatability.

Then in 1686 the English naturalist John Ray broke ranks with tradition and expressed the first doubts about the value of Arum in medicine. Such doubts were not to be generally held for another two hundred years, but here can be seen perhaps the first glimmerings of a more modern medical sensitivity reacting against the robustness and unpredictability of this plant's properties.

In 1710 William Salmon produced a remarkable breakdown of every part of the plant and how each could be used in differing ways, perhaps summing up as a eulogy the entirety of the medical uses to which Arum had been put. The liquid juice of the leaves or root were to be used to make the eyes clear; the powder of the root for coughs, catarrhs and convulsions when made into an electuary or lohoc; the fresh green leaves as a cataplasm or poultice treat wounds and ulcers; the fruit of the berries are considered to be more powerful than both the leaves and roots and can be used for stubborn ulcers, sores and nasal polyps; a tincture of the root, taken morning and evening, brings on the menses, warms the insides against colds, guards against poisons and fevers and eases bowel pain. William Salmon then goes on to describe three further different kinds of tincture of Arum (acid, oily and saline) to distinguish the different maladies for which each is recommended. He ends with a tincture made using distilled water, which is applied as a cosmetic for beautifying the skin or against the poison of serpents or mad dogs. An interesting triad of applications.

This latter tincture, however, has the unfortunate stipulation that it needs to be drunk with mithridate. Mithridate is a recipe for a herbal medicine containing 50–60 (or even more) different plants. It is said to have been invented by King Mithridates in the 1st century BCE to fortify and protect his body from poisoning. It was much sought after, difficult to make and, with the issues of translation which we have already come across, impossible to know which of the great many formulations doing the rounds was the 'correct' one.

Regarding the roots, Dioscorides recommended their use as an aphrodisiac (a long-standing attribute of Arum), advising people to use the roots, boiled or roasted: 'Dranck with wine, it stirs up the vehement desires to conjunction'. The dried and baked root has a warming quality and, when taken as a powder or with honey, is said also to be very beneficial for gout and rheumatic pains.

Post-Linnaeus and his *Species Plantarum* of 1753, the nature of herbals began to change: becoming more scientific and less trusting of earlier truisms. Writings also now reflected the growing separation between medicine and botany and increasingly emphasised either the botanical taxonomy or the medicinal pharmacology of their subjects.

Yet Arum still managed to put in an appearance in this twilight of its herbal heyday. In 1756 John Hill produced his '*British Herbal*', where he described its apparently ongoing and accepted effectiveness in treating rheumatic pains and scurvy, and its ability to restore speech in those paralysed when a bruised sample was placed upon the tongue.

In 1885 in *Hardwicke's Science Gossip*, this was further corroborated from personal experience. It was reported that a Dr Lewis was making 'licking electuaries' from the prepared root. He wrote that this gave a sensation of 'slight warmth, first about the stomach and afterwards in the remoter parts (a Victorian allusion to its aphrodisiac properties, perhaps?); it manifestly promoted perspiration and frequently produced a plentiful sweat. Several obstinate rheumatic pains were removed by this medium'. The same doctor also recommended that people chew the fresh root along with a powder made from the shells of animals, to restore the power of speech in cases of paralysis.

While Dr Lewis' first remedy seems to be harmless enough, one can only imagine the pain caused by eating the roots in the raw state as recommended by the second. Heat and processing of the plant will destroy much of the acrid affect, but taking the plant raw is an entirely different (and not recommended) experience. Perhaps the 'cure' may have restored speech primarily through causing the poor patient to cry out in pain until the histamine effect of the raphide crystals kicked in, rendering the mouth and throat numb and any speech therefore impossible. We will never know.

The same paper at least followed the above remedy with the following comment:

'In these more enlightened days it may possibly be difficult to find persons with sufficient faith to try for themselves the truth of the above remedies. Certainly, for my own part, I should prefer, if suffering from rheumatism, a course of our own thermal waters. I need hardly say that this plant has ceased to be used in medical practice.'

So by now, John Ray's first doubts about Arum's medicinal value had come to pass and Arum had fallen largely into disuse – but not entirely. Arum had one further and unexpected trick up its sleeve.

One of the most interesting and remarkable medicinal properties attributed to Arum is its ability to cure neuralgia. This was reported as early as 1897 by a William Fernie. A patented drug called Tonga had begun to be used to treat this condition and it was derived from a plant from the Fijian islands. As this was unsurprisingly difficult to obtain and patented, the cure was expensive. When it was discovered that the plant used was an aroid, a number of physicians tried using the freshly expressed juice of the English Arum to cure the same. It was found that a teaspoonful cured neuralgia of the head and face as effectively as did the more expensive and patented remedy. According to Cecil Prime, however, this was apparently never followed up for no further records can be traced. An early case of suppression of research by a drug company with a patent to protect?

The same writer also described a use of Arum which seems to have been derived entirely from homeopathic principles, though this is never mentioned overtly by the author. Fernie states that the juice of the fresh root causes an 'acrid burning of the mouth and throat' and that the leaves 'when applied externally ... will blister it'. From this, he deduces that 'accordingly, a tincture made from the plant and its root proves curative in diluted doses for a chronic sore throat and vocal hoarseness, such as it often known as 'clergyman's sore throat''. A clear case of 'like curing like'.

This is one of the last mentions of original commentary on using Arum medicinally. From around the mid-1800s it began to disappear from contemporary medical books as the disciplines of botany and medicine really came into their own as separate and distinct disciplines. From this point on, botanical works would increasingly focus on studying the structures, taxonomy and physiology of plants while ignoring any medicinal uses; medical and pharmacological works likewise paid less attention to the botanical aspects of plants, highlighting instead their medicinal properties. It was in this new current of scientific thinking that Arum fell out of popularity: its distinctive appearance alone no longer enough to earn it a place in the literature. This beginning of the modern scientific revolution in thought and approach was effectively the end of the traditional type of herbal, as had been known for much of the previous two thousand years.

In truth, whilst Arum's medicinal virtues are genuine, its irritant and poisonous qualities make ease of use difficult. It may be that by this time, the secrets of how to prepare and use it safely had been forgotten. The new science of pharmacology was not herbalism and people now doubted the knowledge of the ancients compared with that of new sciences. In this changing intellectual current Arum was not, however, completely forgotten. As interest in its medical uses waned, fascination with its wealth of botanical peculiarities blossomed. While it still made an occasional appearance in folk medicine and in the odd herbal, the plant's future lay not in medicine, but in the fields of fashion, industry and scientific study.

Arum is still used occasionally in homeopathy. The root is used to treat headaches which come on in bad weather or from intellectual effort. It is also prescribed for fever and, not surprisingly, for inflammation of the throat and vocal cords when there is a loss of voice. Just those symptoms which one will develop should one be rash enough to chew on a leaf for a little while.

A Note of Warning.
The above material is of course for historical and academic interest only. It is not intended as a guide to using Arum for self-treatment or that of anyone else. In fact, it is not recommended for ordinary mortals to use Arum medicinally at all. In the chapter on poisoning we will learn the full gruesome details of exactly how Arum treats those who use it without due preparation.

Further Reading.

While there exists no single work on the medicinal properties of Arum maculatun alone, the previously recommended titles will mention in passing the properties as discussed in the ancient herbals. Dioscorides' Materia Medica is the main source and its contents are in the main repeated throughout subsequent publications.

Grieve, M. (1931). A Modern Herbal. Grieve gives probably the best summary of all of the properties of Arum in one concise entry. A good place to start.

Luis, L. (2011). Arum maculatum. www.fkog.uu.se/course/essays/arum_maculatum.pdf. Mentions the effect of Arum on skin bacteria.

Hardwicke's Science Gossip: Internet Archive: bit.ly/YM3RhR

Xu, X. (1998). Association of Petrochemical Exposure with Spontaneous Abortion. http://1.usa.gov/XABLa3

CHAPTER 8
A MISCELLANY & THE TRUTH ABOUT BEARS

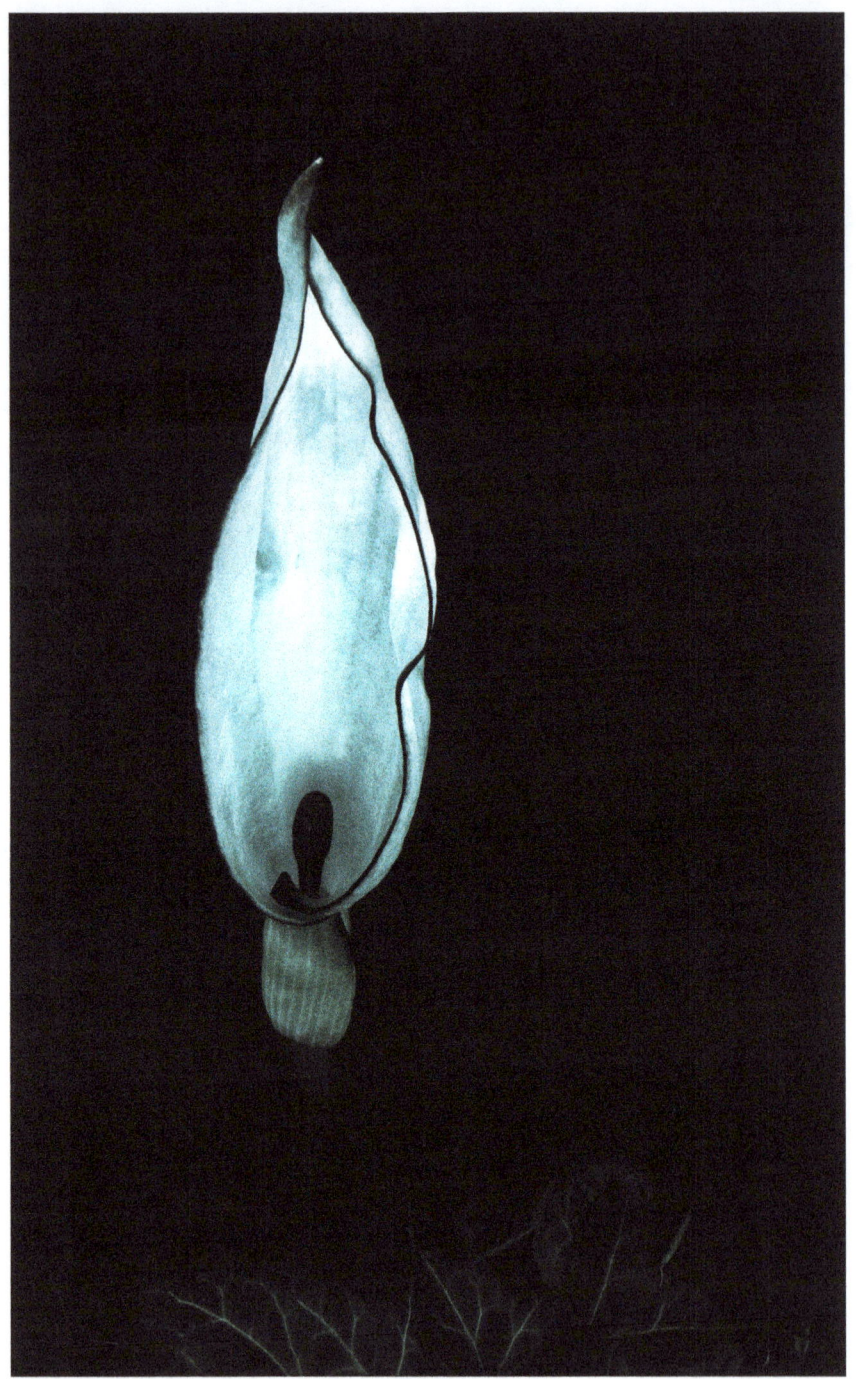

8

Theophrastus started it, around three hundred years before the birth of Christ, though some say he was merely passing on what Aristotle had told him a hundred years earlier. Then Pliny put it into his encyclopaedia and everyone knew about it. By the time that Dioscorides wrote his magnus opus, it was common knowledge that the first thing that bears did when waking from hibernation was to eat up all the Arum that they could find. Gerard told his readers about it and even backed it up with the authorities from whom he was quoting, lest anyone not believe him:

'Beares after they have lien in their dens forty daies without any manner of sustenance, but what they get with licking and sucking their owne feet, doe as soone as they come forth eat the herbe Cuckow-pint, through the windie nature thereof the hungry gut is opened and madde fit once againe to receive sustenance: for by abstaining from food for so long a time it is quite shut up, as Aristotle, Aelianus, Plutarch, Pliny and others do write.'

So, just like many people, the bears couldn't properly waken from their slumbers until they had their bear coffee. Only then were they properly ready to face the world.

In his *History of Four-Footed Beasts and Serpents and Insects*, Topsell (1658) gave a vivid description of the bears leaving their dens and making straight for the Arum:

'... At that time they come abroad, being in the beginning of May, which is the third month from the spring. The old ones being almost dazzled with the long darkness, coming into light again seem to stagger and reel to and fro, and then for the straightness of their guts, by reason of their long fasting do eat the hearb Arum, commonly called in English Wake Robbin or Calves foot, being of very sharp and tart taste, which enlargeth their guts, and so being recovered, they remain all the time their young are with them, more fierce and cruel then at other times.'

The tale of bears' use of Arum as an ursine espresso to rouse themselves into wakefulness after their long winter hibernation has forever been reported as either unquestioned fact (early herbals) or mere myth and legend (everybody else). Yet astonishingly, this ancient legend turns out to be true. In 1987 an international team of biologists published the results of a 2-year study of bear droppings and feeding habits from the Plitvice Lakes National Park in Yugoslavia (now in Croatia). The results make interesting reading. After analysing the food contents of the bear's droppings, the study reported the following:

'Herbaceous plants, mostly grasses, lords and ladies (*Arum maculatum*), and ferns were the most important components of the bear diet from March to May ... Lords and ladies occurred in 67% of the spring scats ... Plants consumed in spring were high in crude protein ... The only plant food with high nitrogen-free extract in spring was the tuber of lords and ladies and bears began digging for this item early in the season. The most important food items in bear diets are lords and ladies and beechnuts. Lords and ladies is the common ground cover in beech forests during spring and early summer; bears feed heavily upon it after hibernation, when other green plants are not yet available.'

So there we have it: Arum really does revive bears, it being one of the few sources of plant protein available early in the year; and it forms the biggest part of their diet for that first part of the year. This illustrates just how much more closely to nature our ancestors lived compared with how we do today. Bears were undoubtedly much more common across Europe a thousand years ago and the paths of bear and human probably crossed repeatedly. In this tale we see the simple reporting of what our ancestors witnessed – newly woken bears digging up Arum roots. And what more obvious conclusion to draw than they did so to wake themselves up from hibernation; which in a way, they do. All that protein; it's the bear's Sunday morning fry-up after the long lie-in of winter.

There is one further legend of Arum that Topsell also reported. In his account, Arum provides a dual role for the bears. Not only does it revive them upon awakening in the spring from hibernation, but it also serves to put them to sleep again in the autumn:

'And concerning the same Arum, there is a pleasant vulgar tale, whereby some have conceived that Bears eat this herb before their lying secret, and by vertue thereof (without meat, or sense of cold) they pass away the whole winter in sleep.'

Humans can also be affected in the same way:
'There was a certain Cow-herd in the mountains of Helvetia, which coming down a hill with a great cauldron on his back, he saw a bear eating of a root which he had pulled up with his feet; the cow-herd stood still till the bear was gone, and afterward came to the place where the bear had eaten the same, and finding more of the same root, did likewise eat it; he had no sooner tasted thereof, but he had such a desire to sleep, that he could not contain himself, but he must needs lie down in the way and their fell asleep, having covered his head with his cauldron, to keep himself from the vehemency of the cold, and their slept all the winter time without harm, and never rose again till the spring time. Which fable if a man will believe, then doubtless this hearb may cause the bears to be sleepers, not for fourteen days but for fourcore days together.'

The ability of Arum to place people into suspended animation for many months over the winter would be wonderful if it were true and no doubt NASA could think of a few good uses for it. In reality though, I suspect that for many of us the temptation to join the bears and idly sleep the long winter away would be too great to resist. Sadly, this particular bit of the legend isn't supported by any independent research. At least, not so far ...

A Miscellany of Uses.

Alongside the main river of Arum's story there sometimes appear, from time to time, springs of other stories; snippets of histories of forgotten uses to which Arum has been put. They surface for a while, never for long, before disappearing again into the earth of history, leaving no more than a fading glimpse of yet another side to this fascinating plant and an insight into the inventiveness of our ancestors.

Arum was for a time recommended as a cosmetic for making the face beautiful. The *Grete Herball* of 1526, and Parkinson's *Theatrum Botanicum* of 1640, both describe how a fine powder is made from the roots to 'cleanse and scoure the face and to smothe the skynne', which will 'wonderfully blanche the skinne, hiding many deformities'. The French developed this into a well-known cosmetic known as cypress powder, though the association with cypress remains unknown. It was still sold up until the mid-1800s. Interestingly, Arum probably did work for this purpose. We have already learnt that Arum contains cucurbitacins. As well as their overall usefulness to the body, these also act as a very effective skin whitener.

Arum has also been used to make soap and the plant does contain saponins in its roots so it would lend itself to this purpose, though it never really caught on due to other sources that were either more effective or easier to use.

One of the most amusing (if surprising) uses for Arum is for riding oneself of an unwanted guest. This seems to have been originally revealed in 1651 by Coles in his *Adam in Eden*:

> 'The fresh Roots cut small, and mixed with a Sallet, will
> make excellent sport, with a sawcy sharking guest, and drive
> him from his over-much boldness, and so will the Powder of
> the dry Root, strewed upon any dainty bit, that is given him
> to eat: For either way, within a while after the taking it, it
> will so burn and prick his mouth and throat, that he shall not
>
> be able to eat any more, or scarce speak for pain.'

It was still around 40 years later in 1692 when John Parkinson, in his *Theatrum Botanicum*, advised the dried root be sprinkled over a guest's meat:

> 'These receipes are recommended for the unbidden
> unwelcome guest to a man's table because it will so burne
> and prickle his mouthe that he has not be able either to eat a
> more or scarce to speak for paine.'

Clearly, table beggars were a known problem in 17th century England. How popular these writers were as hosts following their publication of this recommendation is unknown.

Donald Watts, in his book *Dictionary of Plant Lore* (2007), describes how the berries were given to children and 'simple people' to play tricks on them, presumably for the sadistic pleasure of watching them when the irritant powers of Arum kicked in.

St Withburga and the Fairy Lamps.

One of the more religious mentions of Arum is the plant's association with St Withburga. Withburga was the youngest of four daughters of the Anglo-Saxon king Anna. When he died in 654, his eldest daughter Ethelreda inherited the Isle of Ely and founded the monastery and abbey there. Withburga meanwhile founded a nunnery and church at Dereham in Norfolk. Typically (for a fairy-tale type of legend) she was the poorest of the four sisters and during the building of her nunnery she did not have enough money to buy food for the workmen. After praying for divine intervention, two young does appeared from the surrounding woodland and allowed her maids to milk them. This sustained the workmen in their labours until a jealous local landowner chased the does away with his hunting dogs.

Tradition does not say whether this was after the church was completed or before and if before, we know not how the poor workmen continued to be fed. Tradition does record though that the landowner's experience of divine retribution was swift in its coming, for shortly afterwards he was thrown from his horse and broke his neck.

No further miracles are recorded until long after Withburga died and had been buried in her church. Fifty years had passed without event when her body was discovered to be as fresh as when it was first interred. She is even reported to have blushed when a workman caressed her face with his finger. At this point, events take a dramatic turn.

News of the saint's undecaying body spread quickly and the little church surrounding her fresh remains became a popular destination for the pious. This didn't go down terribly well with the increasingly powerful Bishop of Ely and in 974 he ordered that Withburga's body be forcibly removed and brought to Ely, ostensibly so that she could rest alongside her sisters who were all buried there. Another reason is thought to be so that the Bishop could enjoy the prestige and profit to be made from the increasing numbers of pilgrims journeying to visit St Withburga's miraculously undecaying body. The Bishop of Ely's monks arrived at the church with a cover story of wanting to celebrate St Withburga's presence and miraculous preservation. They plied the locals with food and drink. And more drink. And more still. When everyone had fallen into a drunken stupor (but having somehow kept themselves sober), the monks stole the body of St Withburga and proceeded to take her back to Ely. Part of the journey was along the river Little Ouse. Richard Mabey, in his book *Flora Britannica*, includes a tale from E.M. Porter's *Cambridgeshire*

Customs and Folklore (1969) that describes the part played by Arum in this drama. The monks from Ely rested at Brandon, whereupon nuns from the nearby convent at Thetford arrived and covered the body of the saint in Arum 'flowers' (possibly due to an at the time recognised symbolism with death?). As the boat moved on, these fell from the body and took root in the river banks. Eventually the banks were covered in Arum blossoms that glowed with a pale, white light along the entire length of the river from Thetford to Ely. Meanwhile, back at Dereham, a spring had appeared where Withburga's body had lain. This flows still to this day and has never run dry.

From this legend come the Cambridgeshire names for Arum of Fairy Lamps and Shiners, from how the flowers glow at night. Another name is River Fairy Lamps, said to come from Irish workmen in the 19th century who were brought to the area to drain the fens and discovered along the trail of St Withburga the Arum 'flowers' glowing still, as perhaps they still do to this very day.

Further Reading.

Lidija Cicnjak et al. Food Habits of Brown Bears in Plitvice Lakes National Park, Yugoslavia. (1987). bit.ly/WAWbxm

Mabey, R. (1996). Flora Britannica. A modern source of ancient folk knowledge on our local plants.

CHAPTER 9
THE NOURISHING ARUM

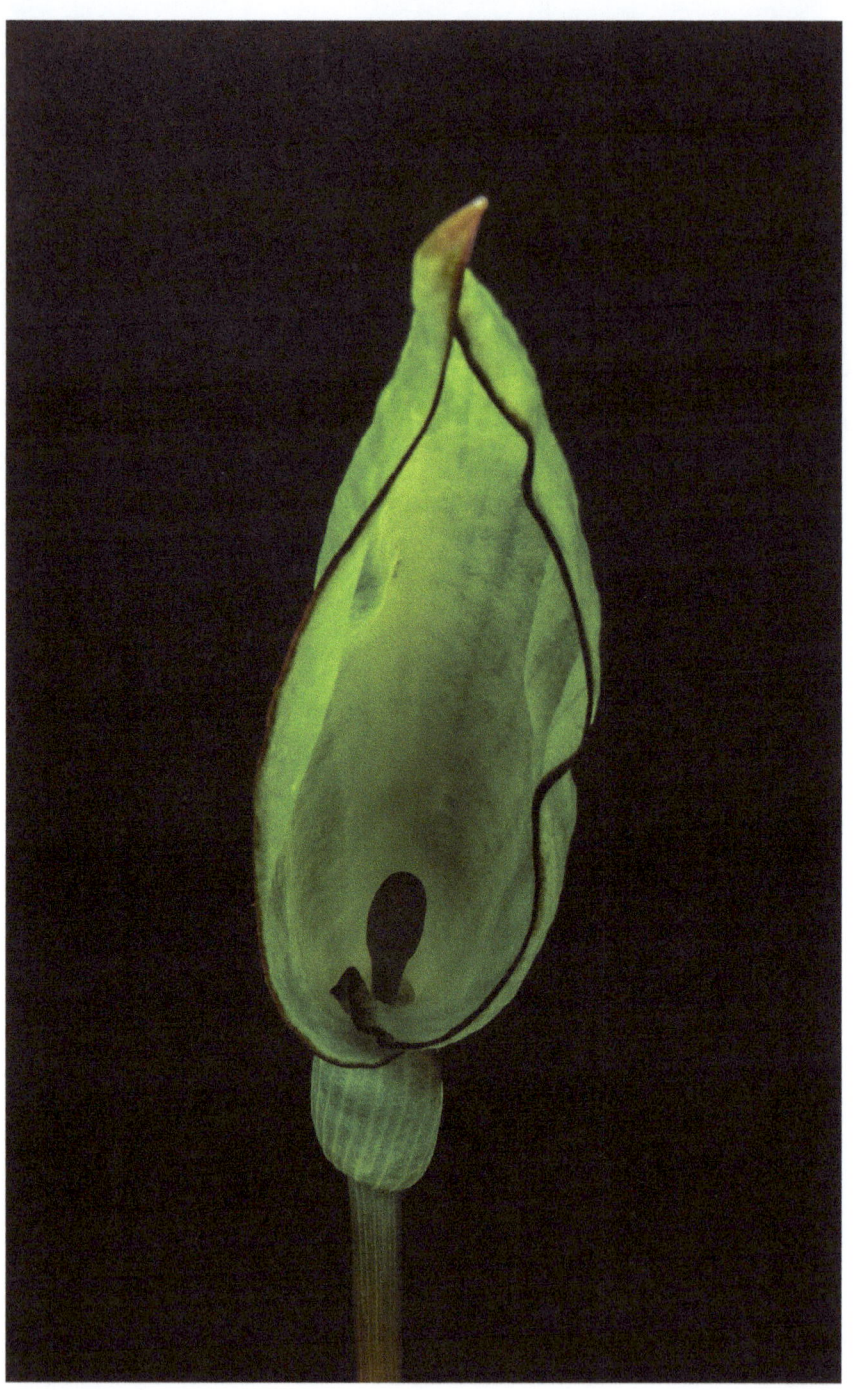

9

W e leave the world of medicine behind us now, putting away those dusty old herbals. It's mealtime. What can we eat? Arum, of course. Once we are sure which Arum.

Arum, as we have seen, is a tricky being. When gathering for medicinal use, we can tumble into our basket an eclectic and carefree collection of different varieties, just as the ancient herbalists did. When it comes to eating however, we need to be more picky about the Arum we are picking. So let us be sure of the Arum in the wood before it becomes the Arum on our plate.

Of the many arums found across the world, certain varieties are genuinely important food crops. Tasty, nourishing and easy to gather. Others are mere substitutes, fall-backs and ne'er-do-wells: technically edible but eaten only in times of famine when the choice is (to paraphrase Eddie Izzard 'Arum or Death'). Arum has a tendency to fall to extremes like that. One could say that it is part of its deepest nature because the crucial factor in this is largely the root and the larger the root the better.

While all arums have nicely plump, starch-rich roots (for very interesting reasons that we will look at later), these vary greatly in size between the different varieties. Take the Egyptian arum, the *Taro*, the 'King of Arums', or perhaps more accurately, the nurturing Mother of Arums. The *Taro* is an anciently important food plant due to its particularly large, starch-rich roots. Gerard, in 1597, described it thus:

> 'There is in Aegypt a kind of arum, which also is to be seene in Africa, and in certain places of lusitania, about rivers and floods, which differeth from that which groweth in England and other parts of Europe. This plant is large and great, and the leaves thereof are greater than those of the water lillie: the roote is thicke and tuberous, and toward the lower end thicker and broader, and may be eaten.'

Immortalised in 1500-year-old carvings at the Temple of Karnak and blessed with oversized, carbohydrate-rich roots just perfect for cooking and sustenance, Taro is undoubtedly the exotic poster child for the edible Arum camp.

Northern Europe, however, is not quite so blessed. As we travel north from the Taro-rich tropical hot spots, the Arum (as we know it) becomes progressively more protective and guarded with its nutritional gifts. While our Arum still has notably starchy roots, reaping this starchiness requires the seeker to overcome a number of off-putting obstacles. In the first instance, the roots of the northern Arums are far smaller than those of its more tropical cousins, giving not nearly so generous a reward for one's digging. That's if you can dig them up'; they are notably difficult to unearth partly due to how deep they like to sit within the earth but also due to their eagerness to detach themselves from the main stem when subject to the slightest disturbance. This is a plant which demands that you take time and care in your unearthing. It requires a stealth dig, an reverent and attentive unearthing as if one were an archaeologist uncovering a valuable artefact rather than a simple 'lifting' as if a gardener armed with the all powerful but unsubtle spade or fork. Finally, in case you have still not been sufficiently dissuaded, the roots are filled with intensely irritating crystals of calcium oxalate which require lengthy processes to render harmless. This is an Arum that demands its payment of hard work before agreeing to release its reward. The Victorians, never afraid of hard work, decided to test this for themselves.

In 1885, *Hardwicke's Science Gossip* published an unusually comprehensive article about Arum. This being the time of Victorian investigation into all hitherto accepted claims of folklore and tradition, the author had tried digging up Arum himself:

'It is no easy task to procure the corms of A. maculatum. Again and again failure marks the attempt to dig them up. I use a fern trowel, but frequently do not go deep enough, with the result that up come the leaf stalks, leaving the corm deep in the earth. The arum loves, too, a soil somewhat stony, and when the plant is met with in such ground, it is well to leave it alone. Anyone who tries to dig up the corm will soon discover the difficulty.'

The *Journal of Ecology* followed this up by reporting that an experiment on Arum had shown that tubers, only 2 cm below the surface in May, had sunk to over 7 cm underground by October. In fact, so fond is the Arum of burrowing into the ground that tubers that were deliberately replanted at the surface had scurried back down to their previous depths within just a week. Arum clearly knows what it likes and this must qualify it for having the fastest-moving roots of any UK plant.

Given that the roots are generally on the small side, have a distinct resistance to being unearthed and are overstocked with sharply pointed crystals requiring a lengthy, arduous and precise process of preparation to disarm them, it is hardly surprising that the northern Arum has never gained the culinary status of its more southerly relatives. Nevertheless, despite these obstacles, it does have a long history of being eaten, albeit often in times of hardship and usually by peasants rather than by royalty. Being one of the few starch-rich plants native to the UK, its roots have often been used to produce a more processed foodstuff, such as flour or to form the basis of a nutritional milk drink, rather than being edible vegetables in their own right.

Arum's culinary qualities are mentioned only as an aside in the majority of the herbals (which, by definition concentrate on the plant's medicinal qualities) but Theophrastus, in his *Enquiry into Plants* in 380 BCE describes the leaves being boiled in vinegar and tasting 'sweet'. He also states how

'men invert the roots of cuckoo-pint before it shoots, and so they become larger by being prevented from pushing through to make a shoot. This way, all of the nourishment is drawn into the root and not wasted in the shoots and leaves'.

He is most likely to be referring to *Arum dracontium* and possibly the Egyptian Arum here, though, rather than the English Lords and Ladies.

Dioscorides describes the berries as having a biting taste (which seems to be putting it rather mildly) and that the plant (presumably the roots) can be eaten roasted with honey. The leaves are described as being used as a vegetable and to wrap around cheese to preserve it. These uses were echoed in almost all later herbals and applied to the English Arum, though whether this was ever based on experience is open to doubt. Dioscorides also states that the 'inhabitants of the Gymnesian Isles [Majorca and Minorca] eat it instead of placentae in their banquets', a somewhat shocking assertion until one realises that, contrary to initial impressions, the inhabitants of these islands were not actually given to eating the afterbirths of newly made babes at parties; the word means simply 'cakes' or small delicacies. Which, though less shocking is still significant; island life must have been pretty harsh if the best the residents could muster for a banquet was a cake made from Arum.

Pliny also mentions the Egyptian Taro, calling it *aron* and noting that it is eaten raw (not something which anyone should ever do with the British Arum, of course). For food, Pliny gives preference to the female plant rather than the male. Sturtevant, in his 1919 book of edible plants of the world, reports that its roots are cooked and eaten in Albania, and in Slavonia (eastern Croatia) it is made into a kind of bread.

Turner and Lyte, in their respective herbals, advise that Arum's roots have to be boiled three or four times until they have lost their acrimony or sharpness, but can then be eaten with meats. This theme is further developed in John Parkinson's *Theatrum Botanicum* of 1640. He gives two recipes for Arum. In one, small pieces of the root are mixed with lettuce and endive, while in the other the dried root is powered and sprinkled over meat. Bearing in mind that this was a time when carbohydrates were a valuable and somewhat scarce foodstuff, this would have added notable nutritional value to a meal.

There is little mention of Arum as a food after this time, aside from elusive mentions of the root being eaten during the Irish potato famine and by the English in times of hardship. The writer Anne Pratt, in her 1870 book describes how she was contacted by a gentleman from Gallway during the famine for help in obtaining flour from Arum. The enquirer commanded a fort in the area around which Arum grew in abundance. He had already tried to eat the plant by roasting and boiling the roots, but they were still too acrid for easy consumption. She advised him to first dry the corns before grating them into water. After steeping for a time they were to be drained and the process repeated before drying the resulting sediment. After carrying out this method, the gentleman wrote that he had prepared 'several packets of flour perfectly free from flavour and fit for use'. This appears to be the last recorded historical use of Arum for food in the UK.

Interestingly, it is still used, albeit on a small scale, in traditional cooking in the Middle East. In Iraq it is eaten by Kurdish tribes who use citric acid, lemon juice or vinegar to neutralise the calcium oxalate crystals. Added to the leaves in a cooking pot, the heat and the acid dissolve the crystals, rendering the leaves harmless to eat. It is eaten as a vegetable on its own or as an addition to *Kubba* (or *Nivik* in Turkish) soup. A recipe for this can be found on the website: ww.sarahmelamed.com and is reproduced with kind permission here.

Kuba (Arum) Soup

Ingredients.
500 g arum leaves
2 med. onions
½ cup coarse bulgur
1 table-spoon tomato paste
½ cup dried plums, seeded
½ cup dried black sour cherries (morello)
1 cup dried apples
½ cup Cornelian cherry (Cornus mas) jam
2 cups water
¼ cup sugar
1 tea-spoon salt
1 table-spoon butter

Method.
Let the dried fruits stand in water overnight.

Boil the arum, strain and chop. Chop onions and sauté in butter for 5 minutes. Add arum and sauté for 5 minutes more. In a separate pan, combine the fruits with the jam and bring to a boil. Add to the arum along with the bulgur, water, tomato paste, salt and sugar, and boil for 15 minutes. Serve warm or cold.'

There is the following note: 'During cooking, do not use a spoon because it will cause the 'prickles' of the arum to appear. Nivik is a poisonous plant, but when boiled and strained the poison is removed. For this reason you must absolutely boil and strain the plant before cooking the dish'.

In the UK, the culinary use of Arum has been reinvestigated in modern times by those with an interest in wild foods and reviving some of the ways in which we used to prepare and use our native edibles in times past.

One such investigation can be found on the countrylovers website (www.countrylovers.co.uk), where there is a vivid description of the modern-day investigation into turning raw Arum into a plethora of foodstuffs such as cookies, tacos, flour and even small bread rolls. With colour photographs showing each stage of the preparation, we can see just what is required, even on a small scale, to extract the starch and to transform this into edible products. This is very helpful for anyone wishing to try out the process for themselves. The man behind this contemporary use of Arum is Marcus Harrison, who runs a wild-food school in Cornwall. The entire process can be seen at www.countrylovers.co.uk/wfs/arum.htm).

A further account of using wild Arum can be found at www.wildmanwildfood.com. The man behind this site is Fergus Drennan, who (among other things) set out to live on nothing but wild food. In 2007 he spent a month doing just that, in preparation for repeating the experiment over a whole year.

His website gives a day-by-day account of his first month. This includes his experiences in preparing Arum flour from the roots and then putting it to use in a wide range of inventive and delicious-sounding recipes, including nut roast and an attempt to make noodles using Arum flour. Three of his recipes are presented here, without alteration, with the third describing the lengthy process of preparation required to render a safe and usable flour from Arum roots.

1. Wild Flour Biscuit Rounds with Creamed Sea Beet and Alexanders Topped with Chanterelles and Fairy Ring Mushrooms.

Ingredients.
For the Biscuits:

5 dessert-spoons Lords and Ladies (Arum) flour (CAUTION: see the warnings on preparing this safely after recipe 3 below)

5 dessert-spoons burdock root flour

5 dessert-spoons tree mallow seed flour

5 dessert-spoons oat flour

5 dessert-spoons reed mace seed-head flour

½ (15 g) of the above flour mix for dusting

Approx. 1 cup sea water

For the creamed sea beet:

8 oz (230 g) sea beet

3 oz (85 g) alexander leaves

2 fl oz (30 ml) oat milk

1 heaped dessert-spoon Lords and Ladies flour

For the fungi:

4 oz (115 g) fairy ring mushrooms

4 oz (115 g) chanterelles

For the biscuits:

Mix the different flours together with the seawater (filtered, boiled and cooled) in a bowl. Knead to form a firm, moist but not sticky, dough. Dust a work surface with a little of the flour mix and roll out the dough to a thickness of 3-5 mm. Cut with a biscuit cutter, place on a baking tray and cook in the oven for 30 minutes at 180°C. Meanwhile, steam together the sea beet and alexanders for 5 minutes before blending in a food processor. In a small pan put 1 heaped dessert-spoon of arum flour. Mix in a little oat milk to form a smooth paste then add another 3 table-spoons of milk. Heat to a simmer, stirring constantly, and then add 4 heaped dessert-spoons of puréed greens. Continue stirring and heat through for 2–3 minutes. While the greens are steaming, 'fry' the chanterelles and fairy ring mushrooms in 2 table-spoons of water (50/50 seawater/spring water).

Place 1 dessert-spoon of the creamed wild greens on top of each biscuit and garnish with the mushrooms. Serve 4–5 biscuits on a plate accompanied by a tumbling pile of mixed chanterelles and fairy ring mushrooms in the middle.

2. Four-layered Nut Roast Served with Dulse and a Wild Rocket and Chickweed Salad.

Ingredients.
For the nut roast:
215 g shelled hazelnuts
290 g sea beet stems
100 g giant puffball
100 g chicken of the woods fungus
130 g marsh samphire
50 g Arum maculatum flour (CAUTION: see the warnings on preparing this safely after recipe 3 below)
20 g dried carrageen seaweed
20 g fresh jelly ear fungus
10 nasturtium leaves
1 table-spoons dried dittander flowers
1 table-spoons natural sea salt
Spring water

For the dulse:
50 g fresh dulse
25 g crow garlic bulbs
10 g crow garlic seeds
Spring water
For the salad:
30 g chickweed
30 g wild rocket
A few wild rocket flowers
1 tea-spoon verjuice
1 tea-spoon wild walnut oil

Method.

For the nut roast:

The Arum here is used as a flour to thicken the nut roast. Rehydrate the carrageen in a saucepan with 2 pints of spring water. Boil for 20 minutes, strain off the seaweed pressing out as much liquid as possible and boil down to leave 8 tea-spoons of liquid.

Layer 1: shell and halve the hazelnuts. Boil these in a little water with ¼ of the sea salt. Once there is only about 2 table-spoons of water left in the pan with the nuts, mix in 2 table-spoons of the carrageen extract and ¼ of the arum flour, the finely chopped nasturtium leaves and dried dittander flowers. Continue cooking on a high heat, stirring all the time, for another 2 minutes. Tip nuts into the bottom of a loaf tin and flatten down.

Layer 2: chop the sea beet stems into inch-long (2.5 cm) pieces. Place in a saucepan with a little water (half covering) and boil for a couple of minutes. Still on a high heat, mix in 2 table-spoons of the carrageen extract, ¼ of the salt and a ¼ of the arum flour. Continue cooking for a further 2 minutes before layering the stems on top of the hazelnuts in the loaf tin. Flatten down.

Layer 3: slice the mushrooms and cook using the same accompanying ingredients and using the same method as for preparing layer 2. Layer 4: prepare and cook as for layers 2-3 except, of course, substituting marsh samphire as the main ingredient. Cover the loaf tin with foil and bake at 200°C for 30 minutes.

For the dulse:

Simmer the dulse in a little spring water for 10 minutes. Add the whole crow garlic bulbs and continue cooking for another 10 minutes. Add in the crow garlic seeds and simmer for a further 2 minutes.

For the salad:

Pick rocket leaves from plant stem and remove any tough lower stems from the chickweed. Mix the leaves together and sprinkle over a little verjuice – extracted from several bunches of under-ripe white grapes – and some wild walnut oil (15 pressed large nuts produce about 1 tea-spoon of oil).

To serve (hot or cold):

Turn out the nut roast and, once cold, cut into slices with a sharp carving knife. Reheat if desired. Serve accompanied with the dressed salad and cooked dulse.

3. Arum Maculatum Shortbread

Ingredients.
4 oz (115 g) butter
2 oz (55 g) caster sugar
3¼ oz (100g) plain soft wheat flour
3 oz (80 g) Arum maculatum flour

Method.
To make the Arum flour:

Wearing rubber/latex gloves to prevent skin irritation, scrub the freshly dug arum tubers clean, liquidise 200–300g batches of chopped root at a time with a litre of water, pour all liquidised root into a 40-litre plastic tub, top up with water and stir thoroughly. This is left for 4 hours or more until the starch and other solids settle at the bottom. With a siphoning tube remove the water from within 2 cm of the sediment, before topping up the whole tub with fresh water. This siphoning and topping-up process is repeated 10 times, after which the sediment is strained through a fine cloth (without squeezing), before final drying in a low oven or food dehydrator. The final solids can then be ground down to a fine flour.

Heat the oven to 190°C/375°F/Gas 5.

Beat the butter and the sugar together until smooth.

Sieve the 2 flours together to mix.

Stir in the mixed flour to get a smooth paste. Turn on to a work surface and gently roll out until the paste is ½" (1 cm) thick.

Cut into fingers and place onto a baking tray. Sprinkle with icing sugar and chill in the fridge for 20 minutes.

Bake in the oven for 15–20 minutes, or until pale golden brown. Set aside to cool on a wire rack.'

Preparing Arum Flour safely.

Fergus gives the following instructions for preparing Arum flour.

'These tubers are extremely toxic, containing very high concentrations of calcium oxalate crystals in the form of microscopic raphides. DO NOT eat any until after lengthy preparation. First, these raphides must be 'leached' out. Strictly speaking, the crystals are barely soluble in water but the method described below separates them from the starch and holds them suspended in water long enough to allow them to be separated and disposed of.

The tubers are best and most easily gathered when the seed-covered shaft is fully grown and the berries are green or red. Dig down 6–12 inches (15–30cm) to find the tubers. Wash well and peel or scrape off all the thin outer skin.

Place in a blender with plenty of water and blitz to as smooth a pulp as possible. Next, pour into a large transparent plastic container. You need to pour in at least 10 times as much water in relation to the amount of pulp you have. Stir with a plastic spoon, cover and leave to settle for 3 hours. Then using a siphoning tube (so as not to disturb the sediment), strain off all the water above the sediment. Add the same quantity of fresh water as before. Repeat this process at least 7 times at 3-hour intervals.

After siphoning off the water for the last time, line a large sieve with a fine cloth (the back of an old shirt is ideal), pour the pulp in and allow to drip dry. Alternatively, once about a third of the water has passed out you can form the piece into a bag and squeeze out the remaining water.

Taste, but do not eat, a very small piece of the pulp.

If, after trying, it leaves a tingling sensation on your tongue, repeat the leaching process several more times before proceeding (this shouldn't happen if you have followed the procedure as described). Lay the cloth and semi-dry pulp out on an oven tray. Place in a low oven with the door ajar and allow to dry. This takes 4-5 hours with an initial 2 kg (over 4 lb) prepared weight of tubers. Halfway through the drying process, mix the flour/pulp together and crumble up any large bits. Once completely dried, grind in a coffee grinder or similar. Two kilos of scrubbed clean tubers yield 400 g of flour. It can be used as an arrowroot substitute.'

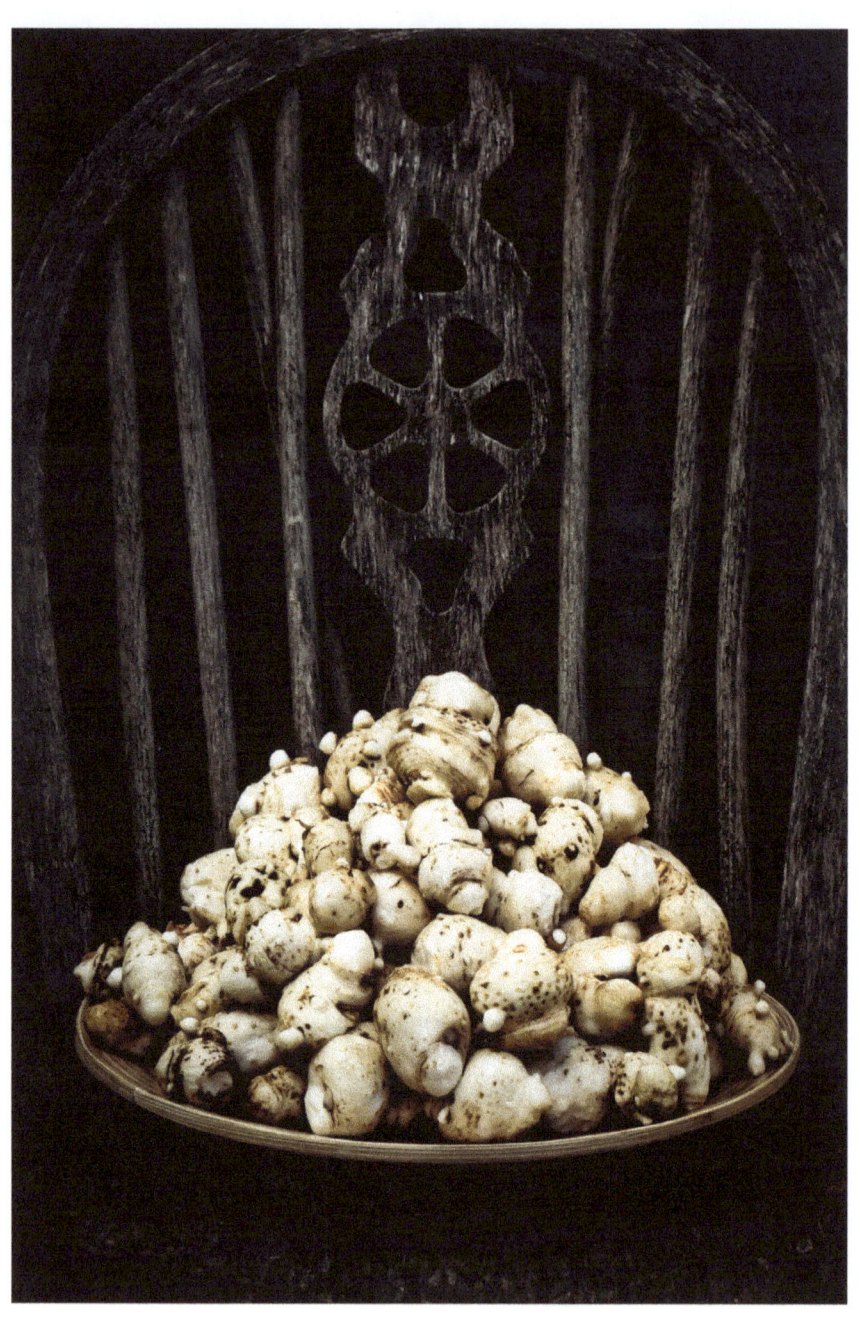

Arum tubers, scrubbed and cleaned. Photo by Fergus Drennan.

Conclusion

It is fascinating to read these contemporary accounts of using Arum as a primary food source. The effort required to gather and process the roots simply to produce only an ingredient, rather than simply picking and eating the plant itself, speaks volumes about our historical relationship with plants, our dependence on them and the lengths we have been required to go to in order to provide food for ourselves.

Just as archaeologists learn much more about past activities by carrying out practical recreations of them, contemporary food enthusiasts such as these are the culinary archaeologists and their experiments teach us much more about the reality of ancient plant use than any amount of reading and research ever can.

Because of the inherent difficulties in using Arum as a food plant, from its irritant and poisonous qualities to the relative smallness of its roots, it has but rarely been sought after as a food source in its own right. It has, however, been much sought after for one particular product which the processes above were extracting: not the flour but its forerunner: starch.

It is this that earned Arum its place in the history books, causing its use to grow to an industrial scale, enabling the evolution of royal fashion and putting a small, out-of-the-way island firmly on the map.

Further Reading.

As with the medicinal recommendations, there exists no single source of information on the edible qualities of Arum alone. For historical information on Arum being used for food, Grieve's Herbal (mentioned above) is a good place to start. For modern information the Internet has two prime sources to assist:

www.bushcraftuk.com This wild food forum is a great place to share and gather information.

www.wildmanwildfood.com Fergus Drennan's website is a source of information also based on direct experience of using wild plants for food.

CHAPTER 10
THE ARUM OF FASHION

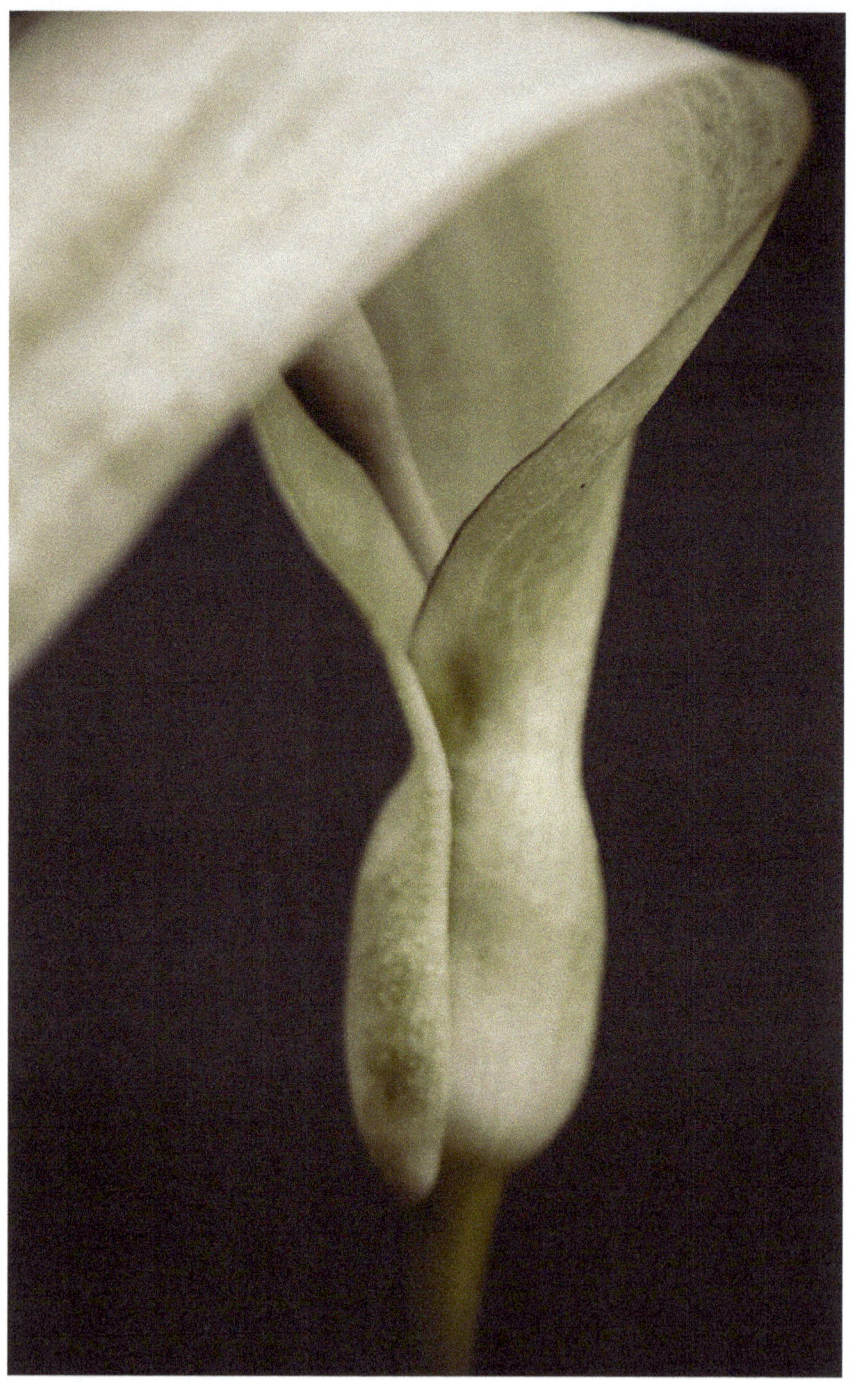

10

The Elizabethan era has already made an appearance as the high point for medical herbals and it presents itself once again for the role Arum played in enabling one of the best-known symbols of Elizabethan high fashion: the ruff.

The ruff evolved from the ruffle. The word and the garment, hand in hand defining the fashion of an era. Ruffles appeared early in the 13th century, as simple strips of crimped trimming added to sheets as a decoration. Fittingly, the word originates from the Germanic *ruffelen*, meaning 'crumpled'. By the 15th century, ruffles were used around the neckline of items of clothing such as chemises, those smock-like precursors to modern-day shirts originally worn by women. They served the distinctly practical purpose of saving a wearer's 'shirt' or top clothing from becoming soiled around the neckline (think how modern shirt collars suffer). Being detachable items they could be washed separately, thus avoiding the need to launder a whole smock or shirt. Very sensible, those Tudors.

With the quickening of the Elizabethan era, ruffles left their utilitarian roots behind them. Hearing the call of destiny, they became ever more ornate and distinctly non-utilitarian items of fashion and status. Bedecked with jewels and arranged into tiers, ruffs (as they were now called), sported varied shapes, numerous styles and a range of colours, all peculiar to the wearer's class and gender. Such were the dictates of Elizabethan society on class and clothing that the yellows, reds and purples favoured by the upper classes are said to have given rise over time to the 'Lords and Ladies' name for Arum. The most extreme ruffs grew to become over a foot (30 cm) wide: supported by internal frames to bear their weight. Such constructions could often comprise over 15 feet (5 m) of cloth. Cecil Prime describes how bearers of such ruffs could hardly walk or bend while wearing them. At one point it was clearly felt to be getting out of hand and Queen Elizabeth banned such extra-wide ruffs. All jolly good fun no doubt, but such experiments in fashion were made possible only by a single discovery – Arum starch.

Starch is the carbohydrate nutrient made by plants as a means of storing energy. When required, plants use enzymes to break down the complex carbohydrates of starch into simple sugars to power new growth. Glycogen, found in our own liver, can be considered the animal equivalent since it performs the same role as an energy store. Starch is found in plants as granules or grains, with the size varying according to the plant species. When extracted and prepared, it is an innocuous, odourless and tasteless white powder. As a prime source of carbohydrates for us, plants with large, starch-rich roots have always been one of the most important elements of our diet, providing both nutrition and energy, and have long been investigated and used for their culinary potential throughout our history.

The use of starch for stiffening cloth has also long been known by humanity, with Pliny mentioning it as far back as AD23. At some point this knowledge reached the Anglo-Saxon kingdoms of the British Isles, though it is difficult to say exactly when and there exists considerable variation in the dates and sources cited. Some clues exist from linguistics. The derivation of the word starch is from the Anglo-Saxon word *stearc*, meaning that which makes strong or stiff. It in turn derives from the root word *starr*, also meaning stiff or rigid (and sometimes applied to corpses). This would seem to give a lengthy precedent to the practice of starching in these lands. One of the earliest written references is from a Norwich church manuscript dating from 1390, which mentions starching but gives no other information. Another fleeting hint is contained in a book from a monastery at Isleworth (1440), wherein is described the practice of using 'sterche made of herbes only'.

Despite these early references, the watershed period for starch making seems to converge on the 1500s, when existing techniques and practices were reinvigorated - or perhaps supplanted - by a Mistress Dinghem van der Plassen from Flemming, a person described as a professional 'starcher'. So impressed were people with her skills in the laundry department that folk would send their daughters to her door to learn how it was done. She charged £5 to teach them and 20 shillings (£1) just to watch; no doubt a goodly income for the 16th century. In her footsteps followed another Dutch woman, Mrs Guilham, the wife of the Queen's coachman. Her skills with the starch led to her becoming the

'Superintendent of the Royal Laundries', and 'titled ladies' came to have tuition from her. This fascination with the properties of starched cloth was crucial to the appearance, just a few decades later, of the elaborate clothing fashions of the Elizabethan era and it was here that Arum came to prominence.

In the world of laundry stiffening there is starch and there is Starch. While any old plant root containing some level of sugar cells can be processed to yield a measure of rough starch, for fine eating and the most rigid of linen stiffening the best starch is the finest grained starch. And the finest grained starch is that made from wheat, rice and arrowroot, all of which yield a wonderfully white, fine grained and pure starch. Not entirely coincidentally, these are also all important food plants.

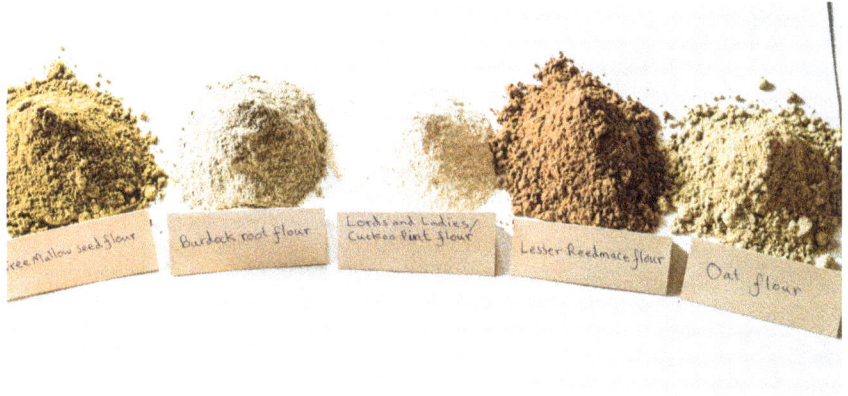

Prepared plant flours. From left to right: Tree Mallow, Burdock root, Arum, Lesser Reedmace and Oat Flour. Photograph by Fergus Drenan.

The problem with using these plants for laundry starch is that they then cannot be used for food production and this was the unfortunate choice which confronted our Elizabethan ancestors: satisfyingly stiff sheets or satisfyingly full bellies. Wheat and corn were the main native starch plants at the time, but they were commonly in short supply and primarily needed for food. The desire to fulfil these competing desires led to the investigation of alternative sources of starch. We do not know how or when, but at some point it was realised that Arum was not only a source of starch but that it produced such a very fine-grained and white starch that it was on a par with that derived from rice and wheat and, indeed, from arrowroot itself. Arum quickly became prized for its suitability for laundry starching, giving a strong and stiff structure to the cloth. So pronounced were its stiffening qualities that, combined with the newly imported Dutch starching techniques, anything seemed possible. People began experimenting. Ruffles seemed a good place to start: thus was born the ruff. The fashion designers of Elizabethan times embraced Arum wholeheartedly and it is no exaggeration to say that it is Arum that made the whole wonderfully outlandish extravagance of Elizabethan ruffs possible.

The Elizabethan era was to fashion what the Pre-Cambrian era was to evolution: a time when a whole variety of strange forms came into being, had their fun and then promptly died out, leaving only those more practical designs to pass on down the ages. Ruffs are the trilobites of fashion: widespread, well known and generally much loved by current generations. They are the 'type fashion' of the era and, like trilobites, now completely extinct.

Less well known is the fact that the Elizabeth era was also the prime time for men to display their creativity in fashion, including that in beard shaping, a somewhat neglected manicure amongst the gentlemen of today. A man wishing to sculpt his beard into a suitably fashionable and attractive shape had to take into account a number of considerations. Firstly there were the Elizabethan class dictates, so that where he lived and what profession he held played an important role in how his beard might be manipulated. Such important social signals were not to be overlooked. That decided, he was then free to enter the constantly changing fashion race of sculpting his beard into different angles, shapes, cuts, coils and contortions.

To ensure such facial architecture would remain in place, men would stiffen their beards with starch. Now, there is a persistent rumour that it was specifically Arum starch that was used for this. The source of this is usually Cecil Prime, who refers to a preface by Thomas Nashe in Robert Greene's *Menaphon* of 1589. The actual sentence is this:

'Sufficeth them to bodge up a blank verse with ifs and ands, and otherwhile, for recreation after their candle-stuff, having starched their beards most curiously, to make a peripatetical path into the inner parts of the City, and spend two or three hours in turning over French dowdy, where they attract more infection in one minute than they can do eloquence all days of their life by conversing with any authors of like argument'.

Robert Green was, along with Marlowe, one of the most established playwrights and dramatist of the time, and was known for writing very popular love tales (the Elizabethan equivalent of Mills and Boon romance) and for living a life of reckless debauchery and continual criticism of others around him – so much so that even he confessed that he was 'unable to keep a friend'. There was a notable contrast between his life and the dreamy, happy romantic world of his writings, but he was acknowledged as a genuinely talented writer and his works were undeniably popular. *Menaphon* in particular was the most accomplished of his 'romance' novels.

Thomas Nashe, on the other hand, was relatively unknown at the time but was part of a contemporary circle of London-based writers known as the University Wits (a group of which both Greene and Marlowe were members). A Cambridge graduate, he has been described as a 'journalist born out of time ... with a brilliant and picturesque style'[1]. He wrote with a fierce satire of the figures and society around him. His preface to Greene's *Menaphon* gave him his big break and he used it to put down Greene's critics and ridicule other writers of the contemporary scene. In the quotation above, he is actually referring to lawyers who leave the legal profession to become writers, saying that they will kill even the best writing stone dead. He then goes on to say how they starch their beards before heading into town to sleep with prostitutes and return with venereal diseases. This gives a good idea of Nash's style, both in his writing and his approach to life. Some have said that this was a coded reference to the young Shakespeare (particularly as Greene himself had given Shakespeare a hard time on occasion). However, we know little of Shakespeare's early life and there is no evidence of him having worked as a lawyer previous to his life as a playwright.

Despite the wealth of information about and by Nashe, he does not explicitly state anywhere that it was specifically Arum that was used as a beard stiffener. We can only assume that it was, given its popularity and penetration into the lives of Elizabethans at that time and their fascination

---

[1] *NNDB website entry on Thomas Nashe. Found at: http://bit.ly/10KyGIK*

with exploring the limits of its stiffening properties.

As an aside, such was the zeitgeist of Arum's starch-stiffening fame that at one time, according to Boyce, a relative of Arum (alpinium) was used in Scandinavia specifically to stiffen clerical collars. As a result, he says, it can often still be found growing around old church sites.

Demand for Arum was considerable for a time, but its super-starch strength came with a significant cost due to its effect on people's skin – not the skin of the lords and ladies, but the skin of their washerwomen. Unlike rice or wheat, Arum contains the incessantly irritant calcium oxalate crystals. These needle-shaped crystals are of such tiny size that they penetrate into the very pores of the skin whereupon they cause intense itching, chapping and blistering. They are equally damaging whether one is handling the leaves or the roots, and Arum starch soon became renowned for causing severe and painful blistering to the hands of the laundry maids who made it. Gerard stated, 'It choppeth, blistereth and maketh the hands (of the poor launderesses), rough and rugged and withall smarting'.

This wasn't the only objection. Some took a dislike to ruffs because they believed that such items were the direct work of the devil. A now infamous pamphlet was distributed at the time by a puritan Christian called Philip Stubbes. It makes for amusing reading; he was the original 'Mr Angry' of every local newspaper letters page. In his pamphlet he gives his views on everything that annoys him, with ruffs singled out for special treatment. 'The devil, in the fullness of his malice invested these ruffes' he said, before going on the describe this new-fangled substance they call starch (another invention of the devil): 'The one anchor piller wherby his kingdome of great ruffes is underpropped, is a certaine kinde of liquide matter which they call Starch, wherin the devill hath willed them to wash and dive his ruffes wel, which when they be dry, wil then stand stiffe and inflexible about their necks'. That was just the preamble. Ruffs (amongst many other things) really did get Mr Stubbes worked up into quite a stiff lather: 'Ruffs that go flip-flap in the wind, and lie on men's shoulders like the dish-clout of a slut'. And so it went on. Oh Arum – in trouble again. The devil's plant, indeed. Philip Stubbes need not have worried. By the end of the Elizabethan era and the new reign of James I, the use of Arum – along with the giant ruffs it had once supported – shrunk and diminished into the realms of history. Laundry maids across the country sighed with relief and Arum was largely forgotten.

Though the love of stiffened ruffs and pointed beards died away, the demand for starch itself continued to grow. A starch-maker's guild appeared, grew into a company and gained and lost its Royal Charter several times. Starch-making became notorious for controversy, competition and conflict. There were Royal Commission fees to pay, competition from starch makers who did not recognise the official guild or company, conflict between the makers and the sellers and, most crucially, repeated controversy over the use of wheat and corn for the making of the starch. Such was the importance of this product that independent production of starch was at times outlawed, with commissioners empowered to break into the houses of people suspected of such nefarious activity. Clearly, such a state of confusion and conflict could not be allowed to continue. Cue the appearance of Arum once again.

Further Reading.

There is a wealth of information on the history of starch and Arum's role in it.

For a historical picture, see Auden, H, A. (1874). Starch and Starch Products. Available from the Internet Archive at bit.ly/Yjv24q

For the full diatribe of Philip Stubbes' 'The Anatomy of the Abuses in England', the Internet Archive once again provides: bit.ly/XAAxLS

CHAPTER 11
PORTLAND. THE ISLAND ARUM

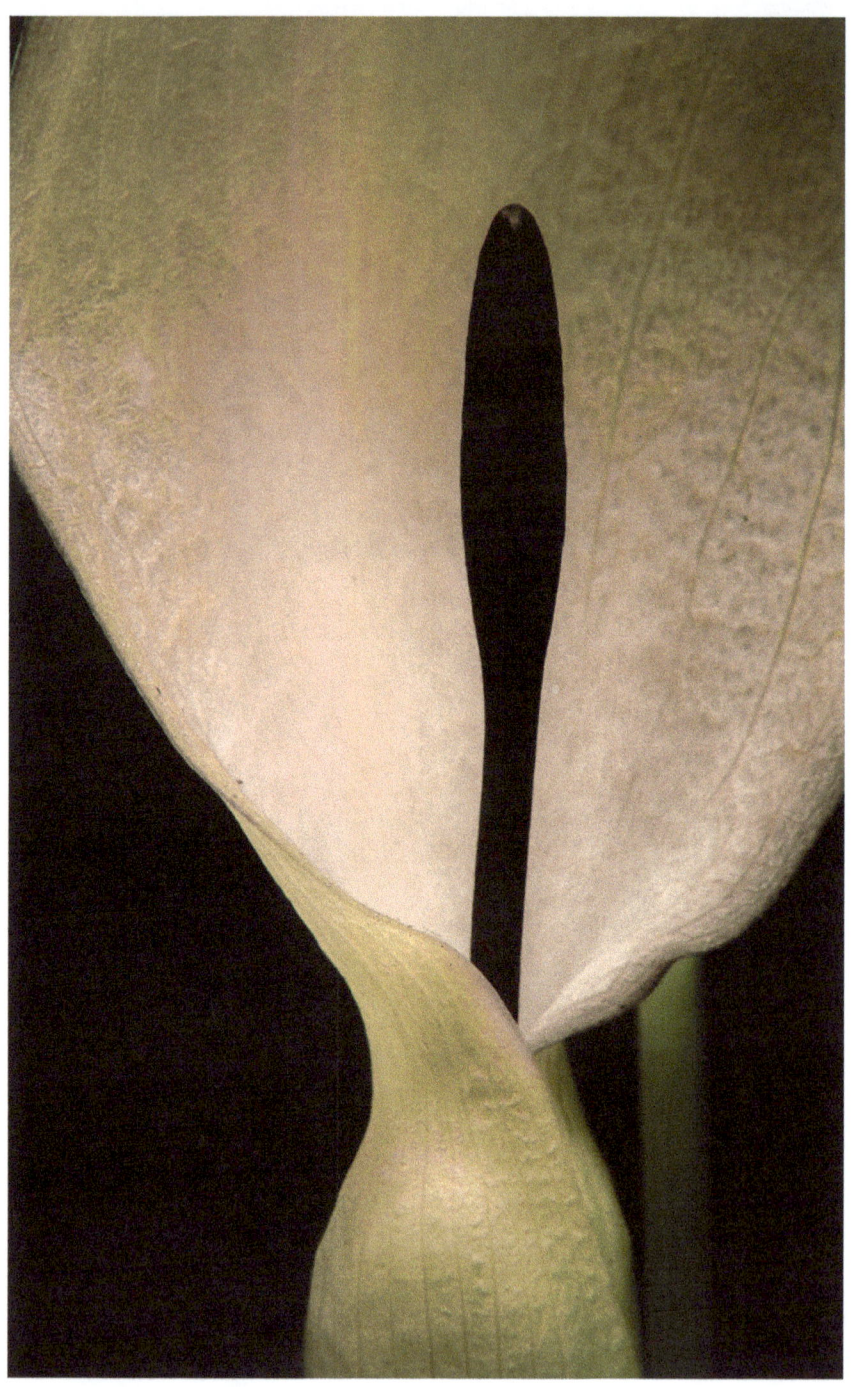

11

A fter lying underground, as it were, for a hundred years or so, in the late 1700s the use of Arum roots for starch sprang into life once again. As before, the shortage of wheat for food was the driving force behind the search for an alternative source of starch. In 1796, the Royal Society of Arts offered a reward to anyone who could find a means of obtaining starch from a source that was not already in use as a food. Wheat was needed for bread, but it was also the main source of starch at the time. Clearly the use of Arum had become largely forgotten since the time of the Elizabethan ruffs. Into the breach strode a Mrs Gibbs from a hitherto little-known island off the coast of Dorset. In doing so, she put the Isle of Portland firmly on the map and initiated a new and influential industry.

Mrs Gibbs.

In her submission to the Royal Society, Mrs Jane Gibbs described the process for turning Arum roots into Starch:

'... A peck of the roots will make about four pounds of starch, though in the operation a less quantity is obtained from some roots than from others: the starch is sold at about elevenpence a pound ... the roots are found in the common fields, and being cleansed and pounded in a stone mortar with water, the whole is then strained, and after settling, the water being poured off, the starch remains at the bottom, which when dried becomes a fine powder. It may be advisable, during the preparing starch from the roots of this plant, to be careful in handling them lest their acrid quality may injure the hands.'

Mrs Gibbs was awarded 30 guineas *(£31.50)* for her winning solution, backed up with a fresh batch of two hundred-weight (metric, around 100 kilos) of Arum starch. The effort involved in producing this quantity of starch from the small corms of Arum is nothing less than heroic – even more so when one of the conditions of the award was that Mrs Gibbs was to produce 'any quantity of the starch whenever required'. That such tasks were carried out hints at the insecurity of food supplies at the time, such that even this effort was preferable to using valuable wheat for starch rather than for food. It also indicates how poor the islanders' lives must have been that an endeavour involving so great an effort was considered worthwhile.

Worthwhile it was though and despite the work involved, this was the inception of a new industry for this most insular of islands and it grew into the most significant and large-scale use of Arum in the history of the British Isles. Interestingly, the Royal Society was initially rather disdainful of Mrs Gibbs' submission, writing that, given the high price of Arum roots, it could hardly become 'an article of commerce or general use'. Yet they grudgingly agreed that Mrs Gibbs had fulfilled the written criteria for the competition because Arum was not in use as a food and she had produced the required weight of starch from it, so they decided to award her the premium of the 30 guineas. Her submission was, furthermore, backed up and certified by no less than three local church wardens, who confirmed that Jane Gibbs 'has in her possession two hundred weight of starch made from roots dug in the common field ...' Perhaps, given their initial low opinion of Mrs Gibbs' solution, they did not hold her to the condition of producing 'any quantity of the starch, whenever required'.

It is quite likely that Mrs Gibbs, in submitting her solution, was promoting an pre-existing activity on the island, albeit probably small-scale, rather than discovering this process anew. It may be, given the isolation of the island, that she was one of the last holders of an almost extinct knowledge of using Arum for starch that had its roots back in the Elizabethan era. It is mentioned in the Royal Archives that starch from Portland was used in the creation of Queen Elizabeth's ruffs. This would date knowledge of the use of Arum for starch by Portland Islanders to the late 1500s – in other words, very soon after the concept of starching arrived in this country. For such an otherwise little-known island, the local population seemed to have capitalised very quickly indeed on what resources they had; 150 or so years later it was no different.

Portland Sago.

The gold standard for starch is (as we already know), that derived from arrowroot. Arrowroot is a South American plant that yields a fine white starch much used in the food industry. It has a neutral flavour, tolerates freezing and comes as a fine white powder. So pervasive has been its use that the plant's name quickly became synonymous with the product itself, in much the same way that items such as Sellotape and Hoover have today. Of all the alternatives, only Arum produced an equally fine-grained, odourless and clean white starch. Because the Isle of Portland was the centre of production, its Arum-derived starch quickly became known simply as 'Portland arrowroot'. The prepared starch powder was transported to Weymouth where it was sold in great quantities, not just for linen starching but also for the manufacture of a health-giving drink for invalids known, appropriately enough, as Portland sago.

Sago is another word for the drink called *salep* (or sometimes *salop*, though that is also a word of insult in French, so we'll stick to *salep* for now). *Salep* is a Middle Eastern drink made from the roots of a certain orchid (*Orchis mascula*), which grows across the hills of Turkey. It has long been drunk for its medicinal and nutritive properties, being prepared by boiling the dried roots and adding milk. It was said at one time that an ounce mixed with animal glue to make a thick soup would sustain a man for a whole day. Delicious, no doubt.

In Turkey, around 20 tons of *salep* is produced every year, requiring 30–40 million orchid bulbs to be unearthed in the process. The resulting scarcity of orchids means that export is now banned and efforts are under way to try to manage the declining orchid populations. In 1855, Rhind wrote about *salep* in his book on the vegetable kingdom. He recommends Oxfordshire orchids as the very best substitute for the true Turkish drink but also recognises Arum as a known alternative and, tellingly, writes that to remove the acrimonious properties of the roots they must be dried and heated rather than just boiled. There speaks the voice of true practical experience. Many writers, copying blindly from earlier sources, have repeated the myth that simply boiling Arum roots is enough to remove the bitterness. It is not. True deactivation of the acrimonious substances in Arum requires sustained dry heat, and only this will render the roots safely edible. The boiling mostly enables a physical separation of the crystals into the water which is then disposed of.

For a time, Portland sago became as popular in Britain (and particularly in London), as true *salep* is in Turkey. Before tea and coffee became the default drinks, it was especially favoured by the working classes, presumably for both its nutritive qualities and cheapness. However, the *Gardners Magazine and Register of Rural and Domestic Improvement* of 1826, while recommending the drink for its health-giving properties, could not help adding that it would be 'valuable to the poor, could one persuade them to go to the trouble of producing it'. So perhaps not all of the poor were as enamoured of it as they 'should' have been.

In 1824, a visiting antiquarian by the name of Fido Lunettes described the then extant practice of Arum gathering. Fido describes how the roots are gathered in July or August and, most interestingly, how the main roots are separated from any offsets which are then replanted. It is from this singular comment that the rumour of Arum being cultivated has sprung. While perhaps not actual cultivation as we might understand it, such a practice is certainly careful husbandry and eminently sustainable.

The collected roots were rubbed to remove the skin and then pounded in mortars into a pulp. Following the familiar process, this pulp was then placed into a sieve and rinsed. The remaining solids were returned to the mortar to be pounded again. These were then left to soak in 6 inches (15 cm) of water for 24 hours, before straining off this water and soaking again for another 24 hours. This water in turn was then drained off and the remaining solids left to dry. Interestingly, it is reported that no metal was to be used in the preparation of Arum.

The End of an Era.

Despite the previous high demand for its products, this island industry of starch and *salep* production lasted only for around 50 years or so, but for those few decades it was impressively influential. It was still in evidence in 1838, as illustrated by this article from *The Penny Magazine* from February of that year:

'On all the fallow-fields (and these are numerous, crops being raised only in alternate years) the Cuckoo-pint (Arum Maculatum) grows in unparalleled abundance, and the field is then called a 'starch-moor:' The roots are gathered by the women, the farinaceous matter is extracted, and a fine supply of British arrowroot secured. Much of it is sold in Weymouth, and the produce brought home in clothing. The Society of Arts, by judicious gifts, formerly gave great encouragement to this manufacture in Portland. Harvest-work is exclusively performed by women; and as none but

Portlanders are employed, a comfortable purse is thus secured by many families for winter purposes.'

This was to be the last recorded description of Arum gathering. Less than two decades later, it had disappeared entirely. Anne Pratt, *in The Flowering Plants of Great Britain* (*c.* 1855), tells of a writer who was living near to Portland in 1853. By that time the practice was said to be almost extinct and never seen outside the island. Anne describes the writer as saying that the farming practice on the island had traditionally consisted of cultivating land for one year and leaving it fallow the next. This is an ancient way of farming land, unchanged since Anglo-Saxon times and it is remarkable that it was still being used at this time. Arum would grow in the fields during the fallow years and the islanders had permission to dig up the corms. This became more difficult when the practice of 'crop rotation' finally reached the island; with fields no longer being left fallow each alternate year, Arum quickly became far less abundant.

The writer stated that by 1853 the only inhabitant of the island who still produced Arum starch was a single old woman, her name unknown. Increased quarrying had by now also begun to eat into the common land where Arum was gathered, leaving ever fewer sites where the plant could grow with any abundance. Such changes no doubt also took place within a time of improved food production and increasing supplies of wheat, so that more extreme substitutes were no longer quite so necessary. Arum had made its mark but the world was moving on.

Interestingly, J. Warren in his book *The Island and Royal Manor of Portland* (1940) writes that the mortars and pestles described by Fido could still be found, though by this date no one was still making use of them: 'In gardens and rockeries all over Portland are many old stone or marble mortars, some of great size. They are doubtless those formerly used in this bygone industry'. Sadly, when Cecil Prime wrote his book in 1960, he had clearly tried to find some trace of the Arum industry but without success, and stated that even the pestles and mortars had by now entirely disappeared since the previous observation of them.

Though all traces of this industry are now long gone from the island, it is interesting to note the location of this intense focus of industrial activity. The south coast of Britain, particularly around the Isle of Wight and the Isle of Portland, is home to a slightly different variety of Arum to that occurring in the rest of the country – namely, *Arum italicum*.

One notable difference between the *A. italicum* and *A. maculatum* is that *italicum* has much larger roots than the more common *maculatum*. Having visited the island and seen both varieties growing there, it is likely that it was the former variety that was being used on the Isle of Portland or, perhaps (in reality), a mixture of both. Having the larger *italicum* variety so readily accessible would certainly have made it much easier for those islanders involved in the lengthy and arduous process of turning Arum corms into starch to gather the required quantity of roots.

The plant can now only be found growing in a few shaded spots on the island, though where it does occur it does so in great profusion.

Further Reading.

The original 'minutes' of the Royal Society meeting which discussed Mrs Gibbs' submission is available on Google Books at bit.ly/ZR9vQP

The Penny Magazine of 1838 has a very well known article on Portland starch available at bit.ly/13y6c5H

CHAPTER 12
THE POISON ARUM

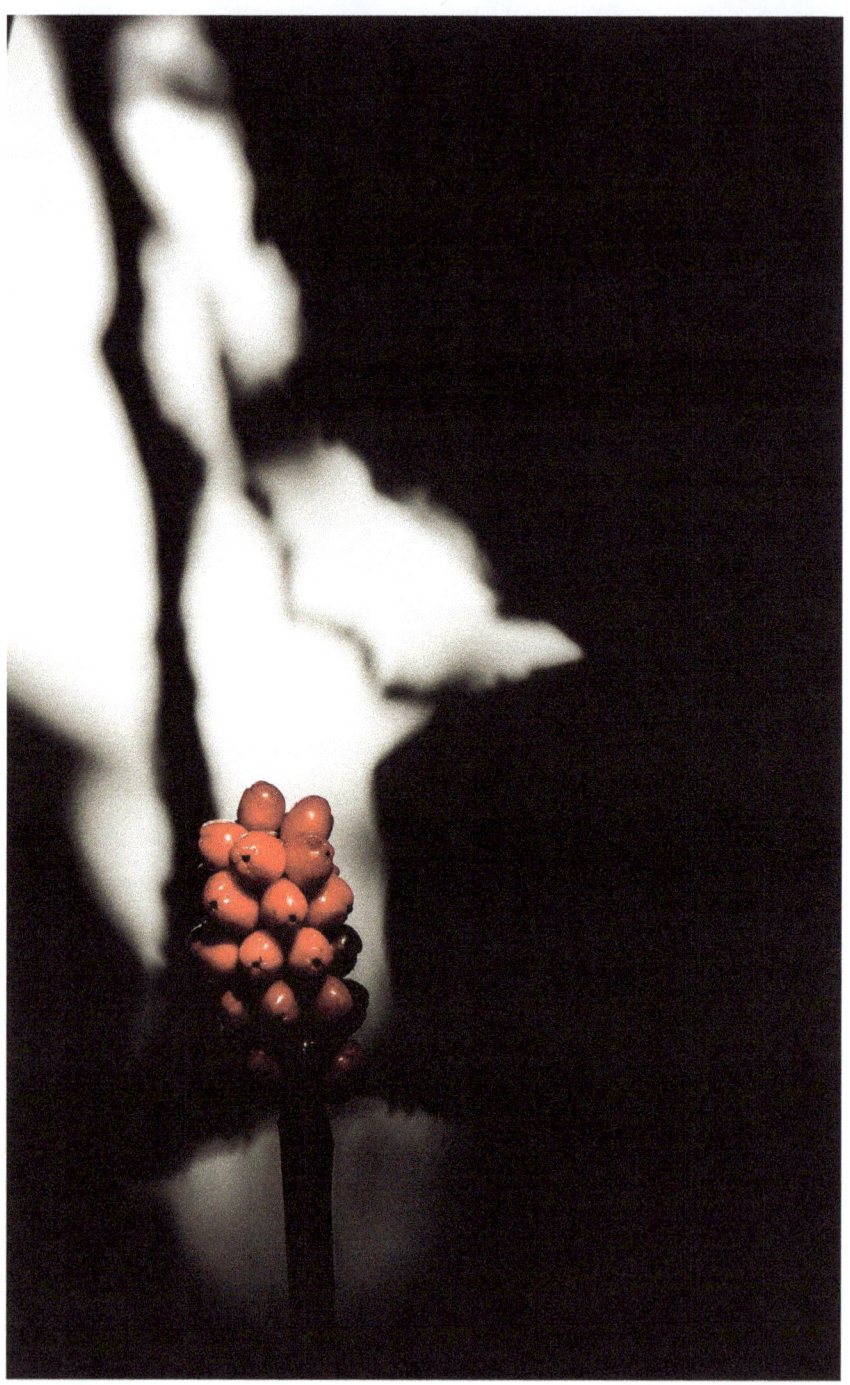

12

If you have recently eaten some Arum and are reading this, I have some bad news for you. For Arum poisoning, there is no antidote. Fortunately, you do need to eat quite a lot of it in order to die. Unfortunately, you would then die rather horribly. So let's treat this as a theoretical exploration rather than a field practical.

Having just journeyed through the history of Arum's culinary importance, it might come as a surprise to find yourself being advised to avoid adding Arum to your list of favourite wild foods, but Arum isn't in any way a beginner-type food. In fact, in many ways it's the plant equivalent of Japanese *fugu* (pufferfish), in that whilst it can be eaten, to do so safely requires an exact knowledge of the correct procedures and plenty of time for the very lengthy processing involved. An easy-eating springtime salad plant Arum is not. As with *fugu*, eating Arum is a matter of dodging its many and unsuspected booby traps and keeping one's fingers crossed for good luck into the bargain. Get it right and you live. Get it wrong and you will experience the consequences, which by any measure, are quite formidable. With this forewarning in mind, let's take a closer look at how Arum views our desires to add it to our evening meal.

As part of their day-to-day metabolism, all plants produce a vast range of chemical compounds for all those tasks which every busy plant has to carry out just to stay in the game and, with a bit of luck, reproduce. While the majority of these substances are fairly innocuous, as with any chemical process there are by-products and some of these by-products are notably toxic, whether to the plant itself, to other plants or to other organisms entirely. Sometimes such poisonous substances are produced quite deliberately precisely because of these toxic qualities. Clearly though, such dangerous substances cannot be left just 'lying around' within the plant's body – they need to be either expelled entirely or stored safely. So, just as we have everything from gun cabinets to underground missile bunkers for our defence machinery, plants have structures called idioblast cells. Idioblasts are where 'normal' plants store their toxic waste. To the casual observer they seem inert, separate from the living, surging fluids of the plant's inner structure and stoically keeping safe their deadly cargo. In Arum, however, these cells are no longer simple storage facilities. They have been quietly converted into the plant equivalent of missile launch silos, each containing a deadly set of weaponry primed for discharge the instant they are triggered – by a trigger such as, perhaps, you thinking that that Arum leaf over there looks like a jolly tasty little snack.

When you bite into your chosen Arum leaf, for the first few seconds you are likely to be blissfully unaware that anything particularly bad is happening. Unfortunately, rather a lot of particularly bad things are happening – you just haven't realised it yet.

The Arum wears beneath its green skin a thin but continuous layer of these highly specialised idioblast cells, effectively wrapping itself about in its own explosive armour. How they work is an example of just how formidable plants can be in defending themselves from attack. In Arum, each idioblast cell is stacked full with plump little bundles of tiny, needle-shaped spears. These spears are known as raphides and each one is made from a long, thin crystal of calcium oxalate. Calcium oxalate is the crystal of oxalic acid, which gives plants like rhubarb their sharp taste. If such crystals are eaten in excess they can cause kidney failure or, at the very least, clump together to form excruciatingly painful kidney stones. But that's not our worry here: Arum has a quite different plan in mind. Each bundle contains up to 200 or so of these crystal missiles with each aimed in the same direction: outwards, ready to be fired into anything that dares to make unwanted contact. Which, as you couldn't resist eating that tasty-looking leaf, you have just done. Oops.

The immediate result of your physical assault causes the cell walls of these missile bunkers to rupture as your teeth bite through the greenery of the leaf. The collapse of the cells' integrity causes an instantaneous expansion of highly specialised mucoid 'muscles' within the idioblasts. These 'muscles' are detonators which break those bundles and power the explosion from each of the dying cells an enfilade of these crystalline spears, shooting out like a rainstorm of tiny shards of glass. When observed under a microscope, this has been described as like the action of an automatic machine gun pumping out round after round of artillery. They travel around 2–3 cell lengths, which at this scale is easily far enough to penetrate the lining of your mouth and digestive tract. Each cell has a 'round' of hundreds of such missiles and each bite of leaf contains thousands of such cells. As Gulliver discovered in Lilliput, they may be tiny but there's an awful lot of them.

Each crystal spear has two grooves along its length, along which are found barbs pointing backwards like the tip of a prehistoric carved fishing spear. These tear into the epidermal layer of your mouth, throat or digestive tract (if swallowed) and physically blast apart the cells there, causing extensive erosion of the surface layers of the soft inner lining of your organs. The barbs come into play here as they prevent any attempt by your body at resealing the damage, ensuring that the physical injury remains open for days or weeks. If this were a gentleman's duel, one might think that at this juncture the point had been made and further attack deemed unnecessary – or at least ungentlemanly. But this is not a gentleman's duel; this is survival and Arum means to win absolutely. Once the physical battering has taken place and your cell walls breached, the real attack, the point of it all, begins. Chemical warfare. Arum's cells don't contain only missiles – they're really just the shock troops. They open the doors for the real takeover, as from each cell there floods a cocktail of neurotoxic and proteinaceous compounds directly into your new open wounds.

Arum now begins to overwhelm your sensitive inner tissues with a variety of toxins and enzymes from the fluid of the now destroyed idioblast cells. Some of these get to work dissolving the proteins in your own cells back into their constituent amino acids. These particular toxins are proteases, similar to those found in snake venom and equally similar in their effect on the body. Which is interesting, given Arum's traditional association as a remedy against snake venom and that it has just used a rather similar mechanism to that of snakes to deliver these substances to your body. Other compounds join forces to send your body's inflammatory response into overdrive while at the same time preventing it from switching off. Still more cause any soft tissue they come into contact with to peel and blister as if burnt. Saponins are found in the root tubers, useful as soap (hence the name and the occasional use of Arum for this purpose), but when taken internally cause the dissolution of your red blood cells. The coup de grace is delivered by the special forces of toxins that physically intensify the pain you are experiencing by stimulating synaptic transmission, giving you the illusion of being in a level of agony and irritation far out of proportion to what might otherwise be expected. The end result is an onslaught of intense physical pain arising from one's very cells, combined with massive swelling and inflammation as one's tissues go into panic mode to fight what is perceived as a widespread internal skinning alive. Bad enough in itself but when it is the inside of your throat that is swelling, well, you can see the potential complications. This is just what is known. The full make-up of Arum's toxic abilities has never been fully analysed. It doubtless holds many surprises still. Best not experiment on yourself just yet.

This retaliation has taken just a few tenths of a second and is repeated anew with each bite that you take. By the time your body realises, many whole long seconds later, that it's in trouble and has transmitted this fact to your brain, it's far too late. The attack is already initiated, implemented and complete.

It's not as if you weren't given fair warning, though. You just weren't paying attention. Like a fortress warding off any would-be attackers by ringing itself in fearsome-looking spikes, the entire surface of Arum is coated with these crystals, all pointing outwards like a bristling porcupine and ready to break off into the flesh of anything that dares to make physical contact. Remember those poor laundry maids and their blistering hands?

It is fortunate that the affects of Arum attack are felt so quickly and so fiercely; it tends to stop people from eating very much of it. That in itself has probably saved many a life. At the physical location of the attack, the symptoms are an immediate swelling and burning sensation, numbness and general pain. More widespread and serious symptoms are experienced if more than a small quantity of the plant is swallowed. As the body reacts to the toxins within itself these symptoms can include convulsions, vomiting, severe purging of the bowels, dilated pupils, insensibility, coma, irregular heartbeat, delirium, difficulty in breathing, obstruction of the airways, paralysis of the mouthparts and tongue, low body temperature, internal bleeding, kidney failure and death. It can take hours, days or weeks for all of the symptoms to subside. Or they don't, and you die.

The earliest documented case of Arum poisoning seems to come from 19th century France and is described in *Flora Medica* (1829) by George Spratt. It concerns the tale of a woodman's three children, all of whom had eaten the leaves of Arum. They were brought to a doctor suffering from 'horrible convulsions' and acute swelling of their mouthparts and throat. Two of the children died in what seems to be an excruciatingly long and painful process, one after 12 days and the other after 16 days. The third child survived after being given milk and olive oil but only because the swelling was, in this case, slightly less severe so that the child was able to swallow these fluids. After a bout of diarrhoea the child's condition began to improve, a common scenario in those recovering from Arum poisoning.

In 1861 the *British Medical Journal* carried three reports of Arum poisoning, again all involving children. The first involved a 6-year-old boy found by his mother lying face down on the roadside. The doctor's report describes his condition vividly:

'I noticed that there was great spasmodic action of all the muscles of the body, and bloody frothing at the mouth; the pupils were widely dilated the eyes set and staring, his hands clenched, and the tongue was bitten. There was also a peculiar choking noise in the throat, similar to that heard in persons whilst in a fit of epilepsy. The lips and face were livid, the heart's action very feeble, the pulse weak and intermittent. These convulsive fits lasted each for about five minutes and then ceased for an interval of nearly the same time; towards the end of each fit, the muscles of the face were affected by a peculiar twitching.'

The treatment prescribed was administration of an emetic which brought about 'two or three evacuations of very foetid faecal matter from the bowels'. After this the child began to shout, 'about guns being fired at him or of devils' and his eyes began to respond to light. With swallowing becoming gradually easier, coffee and more purgatives were administered and he was taken to bed. By the next morning he was still very drowsy but was reported to have recovered after a few days of bed rest.

The second case likewise involved a child found by the roadside. This 8-year-old child could walk but apparently could not see where he was going. He was also suffering from fits and spasms. After a similar treatment of emetics, he was reported to have recovered by the next day.

The third case, dating from 20 April 1860, involved a 3-year old-child from Italy who had eaten the roots. The symptoms were severe, initially of burning and tingling in the mouth before falling into a profound torpor. On wakening and being examined a number of hours later, the child was unable to speak and was suffering from pain and internal swelling from the mouth, down the throat and into the stomach area. Such was the degree of swelling that no emetics or soothing agents could be administered due to closure of the throat. The child was clearly experiencing a great deal of pain and difficulty in breathing and died that night from asphyxia, despite all efforts to reduce the swelling.

Cecil Prime states that the *Pharmaceutical Journal* of 1868 carried a report of six children being poisoned by eating Arum berries, thinking that they were 'green peas', which perhaps suggests the unripe berries but Prime does not give us any further information on this. There are also a number of short references here and there to individuals in more recent times inadvertently eating small quantities of the plant and quickly spitting it out and who, although still experiencing burning of the mouth or hands, did not suffer any further harm. Anne Pratt, writing in the mid-1800s in her six-volume *Flowering Plants of Great Britain*, describes a child she knew who had bitten the spadix of an Arum. The child suffered severely inflamed lips and tongue, which were soothed by milk, but the pain did not subside for over an hour.

In contrast to the severe tales of Arum poisoning in historical records, the Arum of modernity seems to have retreated from causing us any great harm. Or perhaps we have retreated from nature. Undoubtedly we lived much closer to nature even a century ago than we do today. Children and adults alike today are unlikely to be found scouring the countryside picking wild greens for food and mistakenly selecting wild Arum instead. Our tastes too have changed and we are today much less used to eating those bitter-tasting foods found in nature than we once were. That we now find such foods more distasteful may explain why in historical cases the individuals concerned always seem to have eaten much more of the plant than those in recent cases. Whatever the reasons, cases of Arum poisoning today are typically much rarer and less serious, causing only local pain and swelling and perhaps some stomach upset. Undoubtedly, modern medical care also plays its part in this.

John Robertson of the Poison Garden, (http://bit.ly/zW4HSJ) reports that between 1996 and 1999 there were 23 cases of Arum poisoning recorded in England (*Arum italicum*, 8; *A. maculatum*, 15).

These are cases where Kew Gardens was involved in identifying the offending plant material, but none were noted as serious. My own attempts to discover the number of incidents of Arum poisoning reported over the last decade have not uncovered any known cases, though this may be largely to do with the way in which such cases are recorded; nevertheless, it is clearly not considered to be a common occurrence.

In 2002 the National Poisons Information Centre in Ireland surveyed its records from 1997 to 2000. It found 502 incidents of plant poisoning, 74% of which involved, as in historical cases, children. The third most common plant involved was *Arum. maculatum.* The most common cause was berries of an unknown origin and the second most frequent cause was Giant Hogweed.

A similar study in 2009 in Italy looked at incidents of plant poisoning from 1995 to 2007 based on records at the Poison Control Centre in Milan. This looked at over 12,000 cases and found just 45 where *Arum maculatum* had been ingested and 34 where it was the culprit. None of these were recorded as causing serious injury.

Animals.

At slightly more risk are domestic animals such as cattle, horses and goats – and sometimes pigs, though they rarely eat enough to cause themselves any serious harm unless they have nothing else to eat. A case is recorded (www.provet.co.uk) where a herd of goats began suffering from symptoms of diarrhoea and abdominal pain, with a number dying following convulsions and spasms. After moving the animals to another field (whereupon all symptoms subsided), a search found numerous Arum plants and evidence that goats had been eating the berries. Horses and cattle have been similarly affected.

Cecil Prime describes in his book (1960) how he experimented on guinea pigs by giving them nothing to eat but Arum. He discovered that they would die of hunger rather than eat it. He gives a list of those animals which will not eat it, which essentially means all mammals. Yet it is likely that deer sometimes eat the springtime leaves and cowls and birds too will take the berries and are no doubt important dispersal agents for this reason. Birds can often have a devastating affect on the Arum plant by destroying the flowering parts in order to reach the insects trapped within, which must be a rich, tempting and fairly easily accessible food source for any local birds who have discovered this. It may well be a local phenomenon because in certain areas Arum grows unhindered whilst in others the woods are filled with broken, dismembered and destroyed cowls such that very few of the Arum plants mature sufficiently to pollinate.

Treatment.

In the 1800s, treatment for Arum poisoning consisted primarily of physically opening the airway if possible and the administration of some kind of emetic, to force the person, usually a child, to expel the ingested material. Castor oil and coffee were usually given, along with milk, to help absorb the calcium oxalate crystals and mitigate their irritant affect. Beyond that, it was a question of time and the innate strength of the patient versus the amount they had ingested.

Should you come across someone who has eaten Arum, the most important action today is to call the emergency services and request immediate assistance. Modern medical treatment largely depends on the amount of Arum that has been eaten and the age of the patient, but consists initially of removing any remaining plant material from the person's mouth, eyes or skin. Blockage of the airway from internal swelling is the prime danger in Arum poisoning, and management of this can potentially mean the difference between the life and death of the patient. Eye and skin exposure is usually treated by copious washing of the area with water. If you are faced with assisting someone who is suffering from Arum poisoning before trained medical services are able to intervene, then it is important to remember to protect yourself from any physical contact with the plant lest you begin suffering from the same symptoms. Even if the symptoms are mild, it is still essential to seek medical attention because treatment with emetics, rehydration or physical intervention of the airway may all be necessary, including a visit to a medical ophthalmologist if any plant material has been in contact with the eye surface.

Chemistry.

The exact nature of Arum poisoning is still unknown. The idioblast cells are known to be filled with raphide crystals of calcium oxalate, something found in many plants including many edible ones such as sorrel and rhubarb. It is thought likely that this is a common means of plant defence, which works by making the plant inedible to would-be predators. Even insects suffer; studies have found that insects eating such raphide-rich material gain far less nutritional value from it than from plants without such crystals. The nature of the toxins in the rest of the cell is complex but is thought to contain glycosidic saponins, which may increase the acridity of the calcium oxalate crystals; a variety of different alkaloids, mostly used by the plant as part of its pollination methods but which affect the central nervous system in mammals; and cyanogenic glycosides, nicotine and oxalic acid. Most research has focused on the thermogenic properties of the plant and its pollination mechanisms rather than on the details of these chemicals in relation to toxicity, so the exact effect and interplay of these substances, not to mention the full list of constituents, has not well investigated to date.

Calcium oxalate itself is highly irritating to the body and is the main cause of kidney stones, depending on one's diet. It is insoluble and even relatively small amounts can cause permanent liver and kidney damage and death.

The crystalline nature of these raphide crystals is the reason behind the strict methods of preparation required to render Arum edible. Boiling does not destroy them because they are insoluble in water. Heating and drying, however, affects the carbohydrates in the plant and essentially 'fixes' or traps the crystals within the starch. Being non-mobile, they are prevented from causing damage when eaten; though for those susceptible to kidney stones, such foods are definitely not to advised anyway. The essential lesson from all of this is, of course, to avoid eating Arum at all: there are far too many alternative edible wild foods out there which are tastier, more nourishing, easier to prepare and completely non-poisonous.

We have so far journeyed from Ancient Greece through the history of herbal medicine, fashion and culinary habits to the chemistry of poisoning. Which is appropriate, as we now continue the biological theme with a look at the botanical science of Arum, a subject that has proved to be a fascinating and rich source of study for hundreds of years and which, as we have seen with the chemistry of Arum toxins, still has much to be discovered.

Further Reading.

John Robertson maintains the website 'The Poison Garden'. The page on Arum at bit.ly/Z7fheU deals with Arum maculatum.

The British Medical Journal of 1861 can be accessed from a number of sites. This is one: http://1.usa.gov/ZRdNHM

Frohne, D. & Pfänder, H.J. (2005). Poisonous Plants: a Handbook for Doctors, Pharmacists, Toxicologists, Biologists and Veterinarians. & Froberg, B. et al. (2007). Plant poisoning. Both of these contain useful information on the mechanism of calcium oxalate poisoning.

Cao, H. (2003). The distribution of calcium oxalate crystals in genus Dieffenbachia Schott. and the relationship between environmental factors and crystal quantity and quality. A very interesting read on the mechanisms of Araceae poisoning.

CHAPTER 13
ARUM IN EVOLUTION

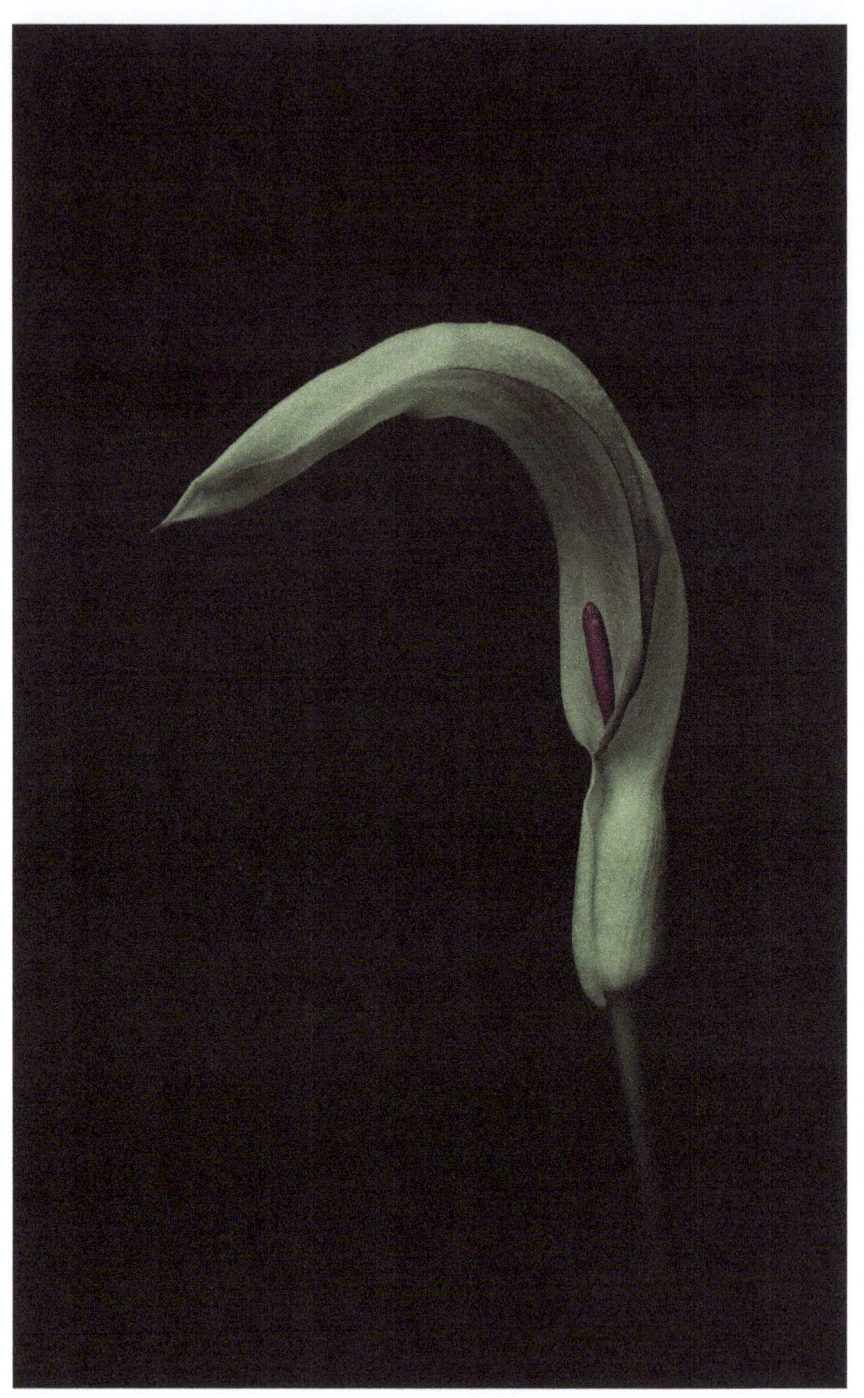

13

'This crafty and antediluvian vegetable'

L et's begin at the beginning. The real beginning. With the tale of evolution and natural selection. In fact, with the evolution of the theory of evolution itself, for Arum played its part in the development of this most controversial of scientific theories, attracting an intelligent, free-thinking scientific and social pioneer on one side and the best minds of Catholic theology the Church could muster on the other. One can readily picture the solidly logical Victorians, with their stout mutton-chop beards, debating this new-found theory of natural selection, filled to the brim with, respectively, Darwinian or Divine certainty on their side.

It all began with a Mr Grant Allen, an Oxford-educated Canadian scientist, writer and social commentator. He wrote on a wide range of subjects, producing material for both academic and popular audiences and many of his works directly or subtly promoted contemporary views on the theory of evolution. Unusually for an academic, he was equally prolific and successful in writing fiction, which he published under his own name as well as under a female pen-name. He even had a bestseller in the form of a racy novel about a woman who refuses to get married and submit to unjust marital laws, is fiercely independent and has a child out of wedlock. All very strong stuff for the 1800s.

His academic works were equally progressive. His book, *The Evolution of the Idea of God* (1897) gives a good indication of the level at which he was looking at the world. It probably didn't do much to endear him to the Catholic Church, but then Allen's father was a Protestant vicar, so he was probably looked on with some suspicion by the Church authorities anyway. In 1881 he wrote *The Evolutionist at Large*, in which he used Arum to illustrate the workings of natural selection and evolution. It was this with which an 1889 paper by the Catholic Truth Society took issue. It really is worth reading both of these works in full; a better example of the early debates about evolution and natural selection is hard to find.

The Catholic Truth Society was founded in 1868 to publish literature on Catholicism. It is still in existence today, publishing material such as prayer books, hymns and prescriptions giving the Catholic view on various societal and scientific happenings. In their newsletter of 1899 they focus their minds on the vexing question of evolution and, in particular, on the statements of Grant Allen regarding Arum and evolution. For Allen, Arum was a true example of natural selection and 'survival of the fittest' in its fiercest form, having evolved a strategy that relied on alternatively attracting and then repelling other organisms around it, to suit its own purposes throughout the various stages of its life cycle. For the church on the other hand, Arum and Allen's statements about it were proof of the absurdity of the theory of evolution.

First to the stand is Allen, explaining each of the unusual characteristics of Arum in turn. The taste of the leaves is so strongly caustic, Allen explained, in order to stop grazing animals from eating them. 'Some early ancestor of the arums must have been liable to constant attacks from rabbits, goats, or other herbivorous animals, and it has adopted this means of repelling their advances', he wrote. Those Arums which were palatable were eaten and did not reproduce, whilst those which tasted less good survived: just as nettles and thistles have done. Allen does admit, however that the 'patient, hungry donkey' will eat whatever is put in front of it. Despite the donkey's seeming refusal to submit to natural selection's dictates, this is left unexplained.[1] .

The reason Arum stores its starch in its roots is to hide it away from small rodents, 'who too frequently appropriate to themselves the food intended by plants for other purposes'. After successfully repelling grazing animals (apart from donkeys) and defending its nutritious riches from pilfering rodents, Arum then switches to attracting organisms, in this case flies, because it uses flies for pollination. To this end it produces a noxious smell to attract them, grows a cunning insect trap to keep them and plies them with sweet nectar to sustain them. All the while showering them in pollen. The process ends when a fly carries in pollen

---

[1]   *However, Allen was certainly on the right track here, as a number of research papers have since shown that calcium oxalate crystals (with which Arum is well armed) do seem to have a role in deterring insects from eating plants which contain them.*

from another Arum and succeeds in achieving pollination. Whereupon Arum's hooded structure withers away, freeing the now un-needed flies to pollinate another Arum with the pollen in which they are now covered.

After this hectic episode, Arum produces its luminous berry cluster and adapts its cunning strategy once again, this time to one of fatal attraction. Its berries are bright red to attract birds but are deliberately poisonous to kill them:

'The robins or small rodents which eat them, attracted by their bright colours and pleasant taste, not only aid in dispersing them, but also die after swallowing them to become huge manure heaps for the growth of the young plant.'

And that is a cunning plan of evolutionary strategy indeed, with the young Arums growing out of the chests of the dead birds like a slow motion version of the 'Ripley alien' we know from the movies.

Now, granted, from our perspective, we can see hints of truth in these ideas as well as areas where the reasoning has perhaps gone astray and since been corrected, but back in the 1880s this was cutting-edge scientific thought; and, in all seriousness, new and complicated ideas – especially those as complex and unintuitive as evolution – always benefit from constant discussion and discourse. Knowledge of exactly how natural selection works was in its infancy back in the 1800s, and even today we undoubtedly know only a sliver of the whole story. The response of the Catholic Truth Society is, on the one hand, perfectly scientifically valid in how it challenged Allen's reasoning and in some ways shows that they were looking at the process of natural selection with a more critical and thoughtful eye than perhaps many scientists at the time. On the other hand, they cannot help but veer off into the realms of anthropomorphic morality judgements when discussing certain aspects of the Arum's pollination strategy.

They first take issue with the problem of how it is that the Arum, a mere plant, can have evolved such a complex strategy yet all the birds and mammals that it manipulates have not yet been able to outwit this 'crafty and malignant antediluvian vegetable', as they so suspiciously describe the Arum. To this they find no satisfactory answer.

They then examine the problem with Arum's described method of pollination. They first of all criticise the flies, who seemingly refuse to learn how they are being manipulated, for after being trapped in one Arum, 'on being released they proceed to plunge straightway--despite experience--into another Arum hood'. They also pick up on the fact that the first Arums to flower must release their flies without having been pollinated themselves – so surely they must die out, leading to an Arum arms race always to be later to flower than any neighbouring flowers.

Their criticism runs deeper still. On cutting open an Arum they discover that the hairs which supposedly are preventing the flies from leaving in fact pose no obstacle to their egress at all:

'The threads are by no means thick set, they twist about and do not run straight, and there is generally plenty of room between their extremities and some portion of the walls. Flies there are generally in plenty, little black flies, so small that it would seem to be a matter of no consequence which way the spikes point, for they could pass between them.'

So why are the flies found in the Arums at all if they can escape whenever they want to? The answer would have appalled their 19th century readers. The flies are actually lazy, no-good drunkards:

'The real obstacle to egress is a condition which looks very much like being drunk and incapable. They lie, often many deep, at the bottom, some without any sign of life, many in a limp and languid condition, much like rioters who have broken into a wine-vault. Whether, when they come forth from their confinement, the fresh air, to which they have been so long unaccustomed, gives them strength and energy to hunt up another Arum before they get rid of their coat of pollen is a question requiring a great deal of very close and clever observation for its solution.'

All of which conjures up a wonderful picture of drunken, hung-over flies being brought back from their stupor by the throwing open of the curtains to let the sunlight and fresh air into the all-night party room, as the Arum's hood withers away and exposes all to the outside world declaring that the revelling is now well and truly over.

Finally, they ask the most obvious of questions: is Arum actually normally found growing out of the skeletons of robins or shrews? At the time, the question was unanswered and it was opened to the investigative zeal of the readers to help discover. This is then followed up by the more difficult question: 'Why are robins and shrews still around if they have not learnt to avoid the deadly poisonous Arum berries?' Which is really a question that just couldn't be fully answered when armed only with a 19th century-level knowledge of evolution. Dismissively, they conclude:

'The plain fact is that the whole thing is too absurd for serious discussion, were it not that so large a number of readers would appear to take such histories for serious contributions to science'. Grant Allen in particular 'proclaims his championship of the crudest and baldest materialism, and his devotion to the creed of 'evolution as a cosmical process, one and continuous from nebula to man, from star to soul, from atom to society'.

The discussion leaves Arum at this point and ventures off into questions regarding the purpose of beauty and, indeed, the existence and purpose of evolution itself. Both of which are still fascinatingly interesting and absolutely valid philosophical questions today and ones which, hand in hand with mathematics, are leading us to uncover depths to evolution and the natural world that reveal Darwin and Wallace to be the before-their-time geniuses they truly were.

Now that this book has evolved to the present point, it's time to take a look at what is currently known about the fascinating biology of Arum. Guaranteed – no dead robins or shrews involved.

Further Reading.

Allen, G. (1881). The Evolutionist at Large, Chatto & Windus. Available at: bit.ly/XLZchM
Gerard, J. (1899). Science and scientists : some papers on natural history. Available at: bit.ly/ZRmamM

CHAPTER 14
MEET THE ARUM FAMILY

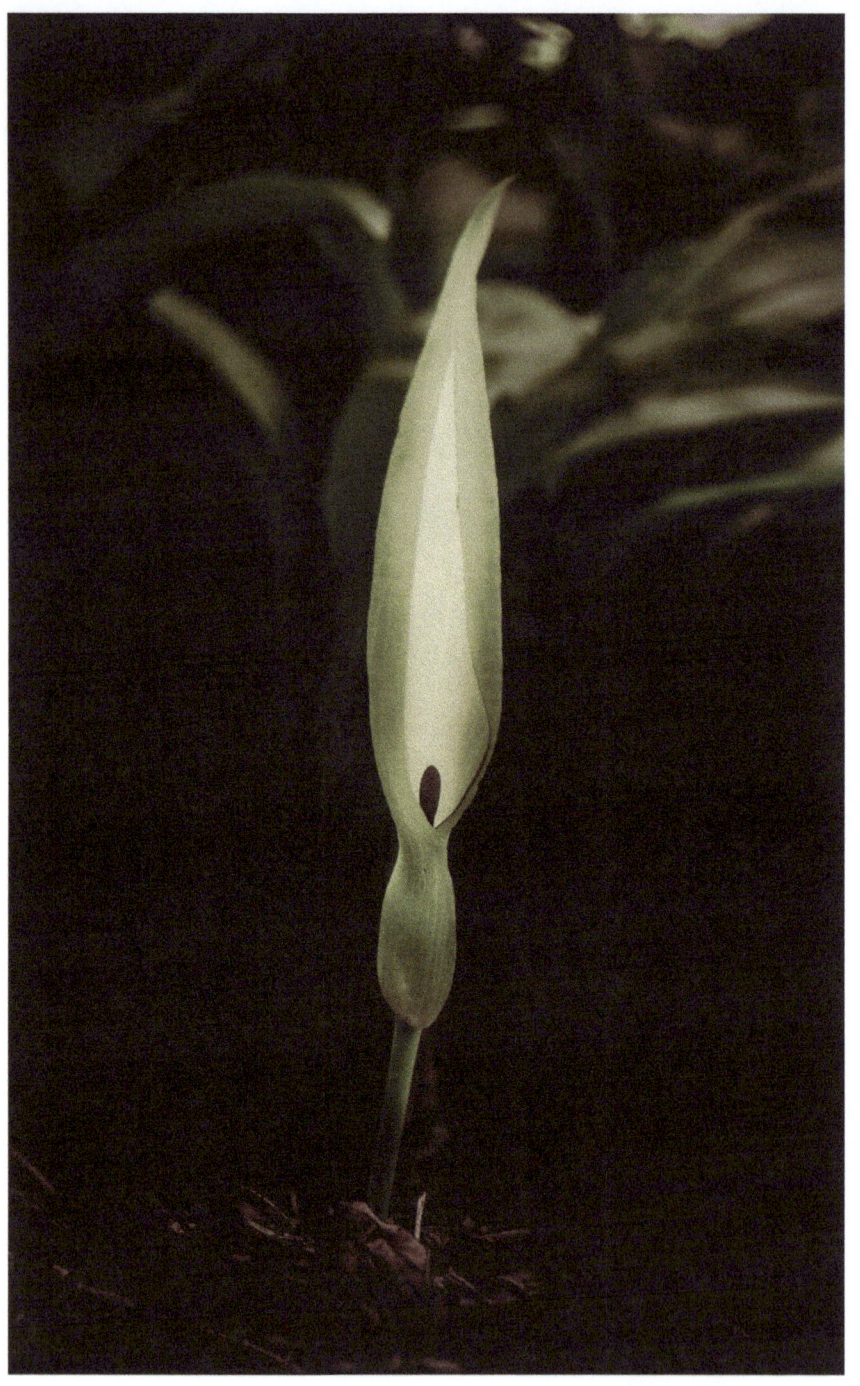

14

T he Aroids are an ancient family who can trace their ancestry back a very long way. Right back to the early Cretaceous in fact, to around 145–65 million years ago. This was a time of huge, warm, shallow seas, when ammonites bobbed about in the water and the widest range of dinosaur species roamed across the land. It is known for the vast deposits of chalk and limestone that were laid down at this time, forming such landmarks as the cliffs of Dover and the prehistoric Ridgeway in Southern England. It is also when the family Araceae first appeared and its member's preference for chalky areas and warm, damp forest ties in well with the likely environment in which this family of plants first evolved. This much at least is certain. It gets much more murky from here on in.

According to modern thinking, the family Araceae (aroids) contains over 109 genera and around 3700 species (depending on how you count them). In the particular genus *Arum* there are currently held to be around 25–28 species. The genus itself is divided into two subgenera – *Arum* and *Gymnomesium*, the latter containing just a single lonely species: *Arum pictum*. The subgenera of *Arum* is further subdivided into two sections and six subsections, which sounds suspiciously like taxonomy-speak for 'I'm not really sure where this goes'. Which is not too far from the truth of the matter.

The traditional taxonomic picture is based on countless hours of close observation of detailed characteristics of many thousands of Arum plants over the past couple of centuries by physical taxonomists, to slowly build up a family tree showing which physical features define each species of Arum. This is a very natural way of classifying the world that mirrors how our own minds work: we look around us and organisms that look the same are taken to be the same species and those which look similar are probably closely related (it's a bit more complicated than that, but that's the general idea). So much for tradition. The limits of this approach have been reached in a number of different fields of classification and the above-mentioned subdivision is an indication of this limit being reached in the field of arum studies. The latest taxonomic research is instead beginning to focus on plants' genetic make-up and examine (among other criteria) the chromosome number to try to define individual species on that basis (again, it's a bit more complicated than that, but that's the general gist). This should be, one would think, pretty decisive but even here Arum pulls us into a mist of Heisenberg-like uncertainty. These genetic-based studies have produced their own species maps, which look quite different to the traditional ones built up over the years, much to the chagrin of traditional taxonomists. It would seem that Arum delights in being notably evasive when we try to pin down exactly where one species ends and another begins.

Part of the problem is that what defines a species depends, much like in quantum physics, on what is observed, so groups of organisms can appear to be one or more species - and more or less closely related - depending on what we decide to measure. The other part of the problem is that arum species in particular display a positively indecent amount of variation even among themselves – a rejection of stability which simply sticks two fingers up at the notion of a distinct 'species'. So the problem becomes one of deciding exactly what to measure because that then defines how we define a species. No wonder the ancient herbals decided to treat them all as a rather generic 'arum' and left it at that.

Lets look at our old friend *Arum maculatum* as a particularly guilty individual. Plants may display either spotted or plain green leaves, while the phallic spadix can be any colour from yellow to red through to purple. Even the shape of the spadix isn't consistent; sometimes it is the same thickness along its length while in other individuals it narrows from a broad base to a rounded point.

Unusually, Arum plants can also be either left or right-handed. In general, plants growing in any kind of spiral fashion, such as climbers like honeysuckle or hop, will do so in the same way: they will all coil either clockwise or anti-clockwise. Even Darwin noticed and commented on this, stating that most plants coil anticlockwise. So all honeysuckle plants, for instance, will all twist in the same direction. Not so Arum. The spathe or 'hood' can be coiled either sinistrally or dextrally (i.e. plants can be left- or right-handed), seemingly at random – a very unusual characteristic indeed to find in a single species. Between 1909 and 1914, a botanist called Sowter spent 5 years examining 1000 Arums he found in Gloucestershire and Essex to determine whether dextrally or sinistrally coiling plants were more common. He found that sinistrally coiled Arum plants outnumbered dextrally coiled examples by about 10%. So if anything, Arum is left-handed. Sinistral of course is derived from the same root-word of sinister, with everything which that implies. Very fitting then that Arum should be our 'anti-clockwise' plant.

Less immediately visible to our eyes are the more subtle variations which Arum displays in its number of stomata. Stomata are the leaf pores through which the plant 'perspires' or 'breathes' (technically, they allow gaseous exchange and water to leave, or not, when conditions are dry). The number on any given area of leaf varies wildly not just across different Arum species but also among them. Such variation just shouldn't be. How can one 'species' exhibit such variation whilst still being a distinct 'species'?

In fact, around 12 different varieties and sub-varieties of *Arum maculatum* alone have been described, which does make a mockery of the very notion of a single, discreet 'species'. Arum is a plant that smears itself across the boundary between one species and another, in a type of vegetative quantum uncertainty, as it hedges its bets and refuses to be pinned down to one thing or another. In actuality, it seems that the Arums around us are indeed still actively evolving, with different forms trying out different strategies and generally existing in a state of complete indecision regarding exactly what they want to be. Which is pretty much what one very recent study found.

A 2010 study took a fresh look at Arum classification and evolution but, unlike traditional taxonomy, it did not examine the usual external characteristics such as leaf shape or flower structure. Instead, it analysed the genetic make-up and differences between a range of European species to determine whether this would reveal more accurately the actual number of species, their relatedness and how early or recently they had evolved and split from one another.

The results make for fascinating reading and are likely to be the beginning of a whole new way of studying Arum's evolution and classification. The genetic results showed that Arum first evolved in the early Miocene, around 20 million years ago, in the area around the Aegean. From here, three distinct phases of expansion occurred; firstly in the late Miocene, followed by another in the Pliocene and the final dispersal happening in the Pleistocene. With each migration, Arum extended its range as new forms evolved and spread out from the ancestral home. The Mediterranean, it appears, is the great cooking pot of Arum in evolution, just as it was for the literature of the ancient herbals in which Arum appeared so frequently and appropriately.

The study found large differences from the standard classification of Arums based on the structure or morphology of the plants. It reported that the Arums found across Europe include plants which, though very different morphologically (i.e. to our eyes), are genetically speaking very closely related. It also found a new species, on the genetic level, which had been missed on the observational one. Even more variety was found in the number of chromosomes exhibited by Arum. Arum exhibits di-, tetra- and hexaploid chromosome numbers, even within what were originally thought to be the same species, which is worrying because the chromosome number is often taken to be a pretty unique characteristic of most species. Most Arums are diploid: that is, they have two sets of chromosomes in each cell, much like ourselves. However, one particular species of Arum has gone down the hexaploidy route, having six sets in each cell, which should see it through most eventualities one would think (and indeed, it is thought to be a more recent evolution). Tetraploidy (three sets) is found fairly commonly and seems to be a popular option, with Arum having evolved this independently on a number of occasions. The chromosome number for Dutch *Arum maculatum* is 28, whereas in England it is 56, with English plants being the younger, more recent variety. Traditionally they are held to be the same species, so one can see the difficulties here.

The study concluded that the only way of separating different species reliably is based on the ploidy or chromosome level. Trying to construct family trees of species using commonly observable characteristics such as leaf shape, etc. is too unreliable because these characteristics keep evolving independently all the time (particularly in Arum) and don't correspond very well to the evolutionary history of Arum. This type of investigation is, however, while not exactly in its infancy, certainly still relatively new and such research is, almost by definition, a bit cutting-edge and pushing at the boundaries of both technique and knowledge.

The long fascination with aroids is evident in the enormous amount of research and interest that this family of plants has attracted. A constantly updated paper, 'History and Current Status of Araceae Research' (which can be found on searching the Internet and is publicly available), summarises the extremely impressive breadth, scope, diversity and complexity of research, from researchers based in almost every country around the globe, into all aspects of this bewitchingly interesting plant family. It is recommended reading and, if nothing else, will convey the amount of work being carried out into finding out what makes this family of plants tick.

In the next section, we will look at the most obvious and immediately interesting aspects of Arum: its pollination story.

Further Reading.

Boyce, P. (1993). The Genus Arum. A Kew Magazine Monograph

Espíndola, Anahí, et al. New insights into the phylogenetics and biogeography of Arum (Araceae): unravelling its evolutionary history. Available at: bit.ly/13ycXVa This reports on a study on the taxonomy of Arums on a genetic level.

Croat, T.B. (1998). History and current status of systematic research with Araceae. Available at: bit.ly/15rfnDj

CHAPTER 15

THE MULTI-ORGASMIC ARUM

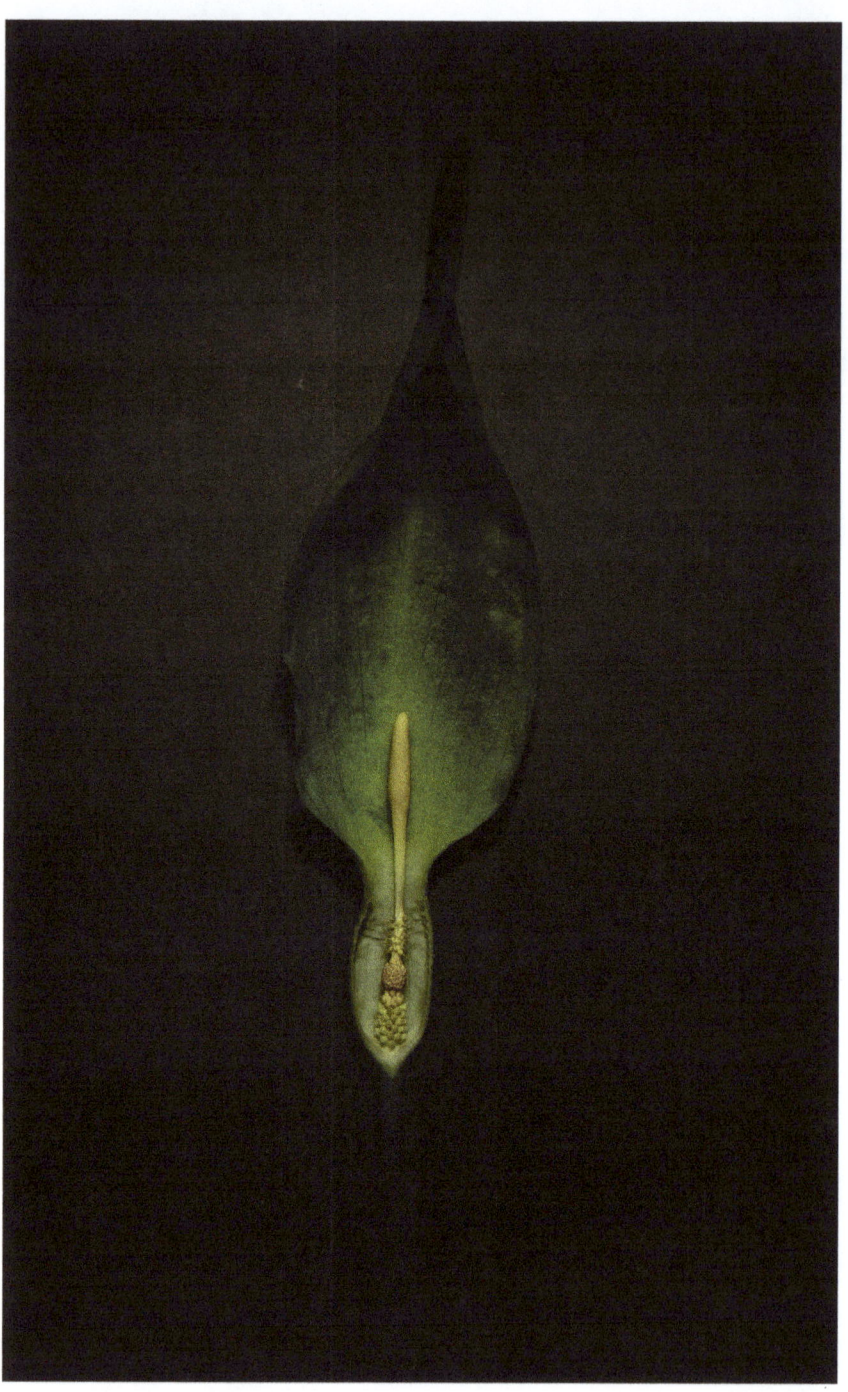

15

Three features of Arum are immediately striking when studying this plant: its unique shape, its habit of trapping flies and its ability to heat itself up to an extraordinary degree. All three of which are aspects of just a single function – how it reproduces. With Arum, of course, it's all about sex.

Hidden from our gaze, the Arums we pass on our country walks are engaged in a brief but intense communal love affair, spending their nights calling out in a language of strange perfumes, using love-smitten flies as pollen-carrying midwives and generally indulging in a process of lovemaking that is so literally hot that the fire of their passion can be felt to the touch. Want to slip beneath the boughs with Arum? Be prepared.

The Curves of Arum.

The story of Arum's complex and sophisticated means of reproduction begins with its salaciously curved body. What we might first think of as Arum's flower really isn't. What we see is mostly just leafage, which hides and protects the actual flowers from the outside world. Yet as ever with Arum, all is not what it seems and in many ways it is more useful to view the entire above-ground structure as one composite reproductive organ, because every part of it is perfectly designed to aid the process of fertilisation. And what a fascinating process it is.

The shapely green structure that makes the plant so photogenic is called the spathe, which curls protectively around the delicate reproductive parts. Technically, it's simply an overgrown leaf bract but it has evolved into much more, developing into a unique and intricate structure dedicated to making the whole pollination mechanism work.

The priapic pintle, so beloved of Arum's folk names, standing within this green cowl is called the spadix and is the most active part of the reproductive machinery. The spadix carries the female flowers at its base, out of sight in the floral chamber created by the same leafage that forms the curvaceous sheath of the spathe. Above the female flowers are the male ones and, above those, a ring of downward-pointing hairs formed from specialised and sterile male flowers. The top part of the spadix that we see when looking at the plant is known as the appendix, though calling it the 'pintle' is perhaps more suggestively appropriate in so many ways.

The structure as a whole is called a 'trap flower', because of how Arum captures its pollinators and then holds them captive in the bulbous base of its floral chamber until they do their duty with pollen carried in from another Arum. It used to be thought that Arum feasted on the insects it captured for extra nutrition. Arum, however, is not the carnivorous type. Instead, the flies Arum captures are well looked after, as befits their status as the sexual go-betweens of Arum's courtship. They are the carriers of fertility between one Arum and the next and around which Arum's entire 24-hour love-fest is arranged. Now that we know our way around an Arum, we can take a closer look at the intricate business of Arum's heated reproduction.

The Fire of Passion.

In the ancient herbals, Arum is portrayed as a notably fiery plant, on account of its taste and the nature of its medicinal action on the body. As we have seen, even its name is based on an ancient word for fire. It is fitting then, that its reproductive process too is one based on fire. And what a fire Arum produces.

It is normally only mammals that are thought of as warm-blooded. The ability to raise one's body temperature above that of the surrounding environment is normally a defining characteristic of advanced organisms. Fish can't do it. Neither can insects. And plants certainly aren't known for it. Well, apart from Arum that is. Arum's strange ability to heat itself was first officially noticed in 1778 by Lamarck in his *Flore Francaise*, but was surely well known long before then unofficially. This ability is formally known as thermogenesis and subsequent studies have continually raised the maximum temperature recorded from an Arum plant, with the highest being an astonishing 32°C. Charmingly, it was initially thought that Arum's warmth helped insects to survive those cool summer nights, acting like an all-night truck-stop with the heaters left on for insects caught short by nightfall. Instead, the heat that Arum produces is purely and simply the fire of its own lovemaking. True to form, when Arum reproduces, it does so with a fiery passion few others can equal.

The process of Arum's self-immolation is under constant research but one of the latest studies has found that Arum experiences four separate surges of heat over the day or so that Arum's lovemaking lasts. Each wave of heat is connected to a particular phase of sexual activity and fertility. Though the main focus of the heating is on the male parts of the plant, the female floral chamber responds with its own waves of heat that pulse in time with those of the male flowers, in a dual blaze of sexual harmony. These hot flushes have been described, even by otherwise restrained and detached botanical writers as 'metabolic explosions' and also, more tellingly, as 'paroxysms'. Both of which sound suspiciously like scientific euphemisms for orgasms, which is really just typical of Arum – our only multi-orgasmic plant.

The first 'paroxysm' of heat occurs the day before the spathe opens. It's a mere flicker of heat compared with what is to come, almost a foreplay to the main action but it is this short-lived warming of the whole plant that is the trigger which ignites the internal flames of the Arum. Dioscorides prophetically warned us that Arum induces an 'insatiable sexual desire', and when we become caught up in the forest fire of Arum's own roller-coaster of multiple orgasms, we can see why. How did the ancients know this?

After the foreplay comes the real fireplay. With a second and prolonged burst of heat Arum sets in motion the priapic uncurling of its green hood to unsheathe the purple finger of its spadix. This graphic display of external turgidity is powered directly by this much larger blaze of internal fire. It is the external signal that this Arum is ready to mate. Only when the spathe is fully erect comes the third – and by far the greatest – explosion of inner flame. This is the peak of Arum's fiery coitus, when its focus is on nothing but reproducing and it literally consumes itself with its own passion.

This third and climactic blaze of heat is focused intensely on the tip of the phallic spadix, which has been recorded as reaching an incredible 19°C above the ambient temperature. At the same time that the Arum is waving its hot poker around in the air, like a glowing rod from a bar fire, the female parts become receptive and begin to glow with their own inner warmth, gently heating the gestation chamber down below: a feminine counterpart to the cocksure display going on above. When the tip of

Arum's finger reaches its peak of heat it begins to release its own unique mix of pheromones, concocted to attract a rather select community of pollinators to come and visit. This phase of sexual readiness and openness continues throughout the night. At its climax the female flowers open themselves to fertilisation and become wet with a thick fluid as they ready themselves to receive the male pollen carried in by the chosen ones of Arum's distinct perfume.

The Power of Starch.

Strangely, the fuel for all this botanical pyrotechnics is starch. Arum, if you remember, is packed full of the stuff – and this is why. Yet starch is really intended as an energy reserve rather than as a fuel in itself and burning starch directly is a startlingly inefficient way to produce energy. So poor in fact that most other plants have abandoned it for more efficient methods. To produce energy from starch it first needs to be converted to a sugar, which takes time. Arum can't wait. Using a different chemical pathway to normal plant respiration, it leaps straight into metabolising its vast reserves of starch directly, a process that yields little in the way of energy but does produce an absolute firestorm of heat, particularly when it is burnt as quickly as Arum burns it. When measured in terms of oxygen uptake, the metabolic rate of Arum during its most climactic phase matches or exceeds that of a hovering hummingbird. Let that sink in. The Arum plant sitting there on the forest floor is firing away at the same rate as a tiny hummingbird in a rainforest sucking nectar from tropical flowers, burning pure sugar to keep itself in the air. Such passion takes its toll. Just as hummingbirds drink their own body weight in nectar each day, Arum consumes itself in this self-induced inferno of activity. The dry weight of an Arum plant in this state of excitement can shrink from a typical 32% down to an incredible 6% over the course of a single night, so fiercely does it consume its starch reserves with its amorous intensity. Lovemaking has often been held to be a good way to lose weight. Arum doesn't need to be told. Around the maypole it goes.

The Scent of Seduction.

The reason Arum gets so hot under the collar stems from its almost baroque strategy for reproduction. Unlike the roses and tulips of this world, Arum does not attract its pollinating insects by producing big, bright, beautiful blossoms. Such audacious petalicious structures are not Arum's style. Instead, Arum has created a far more subtle way of ensuring that the family line is passed on, because its pollinators rely not on sight but on scent. It is worth noting here that some types of arum go about things slightly differently and are known as 'flag species' because they hoist their flowering parts up high on a tall stem to make it obvious to any passing pollinators that there is something interesting here worth investigating. Flag species are going after flies with good eyesight who will notice the Arum trying to catch their attention as they fly past. The British Arum is looked upon as one of the 'cryptic species' (which I think suits it much better), because it hides its flower low down in the undergrowth, out of obvious sight. The arum species that do this are going after pollinators which follow their nose and not their gaze.

Now some arums are happy with just any old fly. So long as it comes with pollen and goes with pollen, that's just fine – job done. The British Arum, however, is incredibly fussy about who it permits to carry its pollen about. It's not a job open to just anyone. It sets out to attract just a single species of owl midge called *Psychoda phalinoides*. Even more specifically, it's just the females that are targeted: over 4000 have been found in the floral chamber of a single Arum plant. Arum is clearly offering something they like. And what owl midges like more than anything is a nice, fresh cowpat. It is here where the females lay their eggs and, in their nightly flights around the countryside, this is the nursery site they are looking for.

As part of its strategy of seduction, Arum produces a highly complex mix of chemicals which it concentrates in the tip of its spadix or finger. In the climactic phase of Arum's amorous throes, Arum gets its so-aptly named 'red hot poker' so hot that these volatile compounds are vaporised by the heat at the tip and off they drift into the night air. These form Arum's invisible aerial net by which it attracts and draws in its pollinators and it is this that requires the Herculean effort of producing such an intensity of heat. Arum becomes the living oil-burner of the forest floor. The hottest love in town. Just for a night. With the specific mix of chemicals Arum gives off, it replicates the smell of the owl midge's nest site so accurately that the broody female midges can't tell the difference between a fresh, ripe cowpat and a fresh, ripe Arum. In they come.

This highly specialised perfume contains at least three specific chemicals that appear to be the unique attractors for this single species of midge only. Such fussiness (which is not unique to the British Arum) means that certain arum species can differentiate themselves from each other even when in close physical proximity, by releasing specific recipes of odours that attract different species of flies. This is a smart strategy. The number of different odours that may potentially be given off by one plant and hence the range of signals it can produce, is practically unlimited. It allows different arum species or subspecies (if we accept that they exist) to exist side by side, all broadcasting on different frequencies, to different audiences. Like a group of perfume houses all in the same city, they have evolved to target specific niche markets, with each releasing its own, distinctive, in-house perfume, as unique in their way as any flower in our vase is visually.

Amusingly, a number of studies have found a difference in odours between the masculine pintle and the female floral chamber, with the former releasing odours of a 'pungent' nature and the latter smelling 'sweetly'. Frogs and snails and puppy dogs' tails, indeed.

In a very real sense, these tiny flies can be considered to be an airborne part of the Arum plant itself. Without them, Arum could not reproduce, at least not sexually, for they are the fetchers and carriers of Arum's germinal seed. Around them Arum has shaped almost every aspect of its being. They are the reason for its fiery lovemaking, the cause of its unique shape, the judges of its body odour and the drivers behind its entire phallic presence. Arum has constructed itself around ensuring that it sows its wild oats, and it is these little owl midges that do the sowing.

A Perfect Trap.

As we all know, it's one thing to attract someone; it's quite another to keep them. Arum, however, is a possessive lover and once it has attracted its flies, it knows all about keeping them. It traps them and holds them captive until it's had its way with them. How it does so is, quite simply, amazing.

It was in 1926 when an Austrian botanist named Fritz Knoll first demonstrated how Arum was structured to successfully trap its flies.

Remember that flies are the absolute masters of grip. Any surface, any angle, they can cling to it. To achieve their uncanny gripping ability, flies' feet have a mixture of tiny claws and what are, in effect, suction cups. If the surface is rough they can dig in with their claws. If it is smooth, they can secure themselves using the same principle as a phone holder uses to stick to your car windscreen. So how does Arum manage to defeat them so easily?

It turns out that the cells on the surface of the spathe (remember, that green cowl we used to think was the 'flower'?) do not have the smooth, flat shape which such surface layer cells typically do. Instead, they are cone shaped and point downwards towards the basal floral chamber. They are also covered in a thin layer of oil, at least during the time of thermogenesis, when the Arum is in full heat. The tiny *Psychoda* midges coming in to land, after being drawn in by the smell of fresh cow dung, find that try as they might, neither their claws nor their suction cups are able to find any purchase on this surface. Like cartoon characters suddenly losing their grip, they slide straight down the walls of the spathe, through the barrier of hairs and into the floral chamber below: trapped in a trap which is perfectly designed to defeat the otherwise undefeatable gripping abilities of these flies. But Arum goes even further than that. Not only are the insects entirely incapable of climbing out again, they are persuaded from even trying by a very unusual feature of the walls of the floral chamber. Knowing how much flies like to fly towards the light, Arum has structured the wall of the lower parts of its chamber to be more slightly more light transparent than the upper parts. So the light comes in just a little more intensely from the bottom than from the top. Even if some flies decide to be clever and hatch a plan to climb their way out, they would naturally be in complete agreement to head for the light. Which in Arum's holding cell means they will head further in and further down. Repeatedly. Those Victorian scientists were right: a 'cunning vegetable', indeed.

Nevertheless, the flies are well looked after during their confinement: sustained in their new, temporary home by the warmth in the floral chamber which it is thought keeps them active and thereby more likely to brush any pollen they have brought with them against the female flowers. The walls of this floral chamber also have another unusual feature up their sleeve to aid the well-being of the flies: air conditioning.

The cells making up the floral chamber contain 'gaps' between their cell walls, which allow more oxygen to diffuse into the chamber than would otherwise be the case, thus saving the crowd of flies from the fate of collective suffocation. Not only are the flies kept warm and well supplied with oxygen, but they may also be fed too. It is possible, though not yet proven, that the flies may be further sustained by the fluid exuded by the female flowers, which is slightly sugary and so may help to keep them sprightly throughout their confinement until the morning, when the male flowers mature and shower them in pollen. The flies have at this point been described, most appropriately, as being 'tarred and feathered', which seems a very good description. Even if the owl midges do not feed on the sweet fluid from the female flowers, it does seem to help the male pollen to stick to them.

The Fruits of Passion.

The final, almost post-coital, phase of heat happens once the female flowers have achieved their goal. Following fertilisation, it is the turn of the male flowers to heat up, like so many tiny spark plugs and begin raining their stores of pollen onto the flies still held within the floral chamber. When suitably covered, the fine hairs which previously helped to prevent escape wither and the cell walls of the spathe lose their oiliness, allowing the flies to escape by climbing up the tunnel and the central spadix into the open air. This marks the end of Arum's heated love fest; its goal achieved, the plant now begins to collapse. As if undergoing an accelerated ageing, turgidity turns to flaccidity. Slipperiness turns to dryness. External displays wither. The flies are left free to make their escape. Upon their exit, they typically dive straight into another Arum to repeat the entire experience, so enjoyable has it been (remember the drunken revellers of Victorian biology). Fertilisation achieved, the Arum plant now withdraws into an apparent sleepiness.

When Arum reappears, with its crown of late summer fruitfulness, it is literally bearing the fruit of its successful springtime coupling. At this point it is the birds it has its sights on, as they are the most important means of dispersal for Arum's seed, being rather partial to the odd berry or two. Little research has yet been done on this aspect of Arum's life cycle, other than compiling lists of birds known to eat the berries, though pigeons have been marked out as important long-distance dispersers. As far as we know, however, Arum does not grow from the dead bodies of the birds which eat it, as Grant Allen once would have had us believe.

Further Reading.

This is, of course, a very brief and somewhat artistic interpretation of the reproductive biology of plants of the genus *Arum*. There is much, much more to it than I have written here. For a more conventional overview of Arum's reproductive processes, I have listed some of the best sources below.

Dormer, K. (1960) *The Truth about Pollination in Arum*. Although written over 50 years ago, this is still a great description of the process of pollination.

Halevy, A. (1985) Arum, in the *CRC Handbook of Flowering*. This paper beautifully describes Arum's thermogenic processes as 'paroxysms'.

CHAPTER 16
ARUM IN ART & LITERATURE

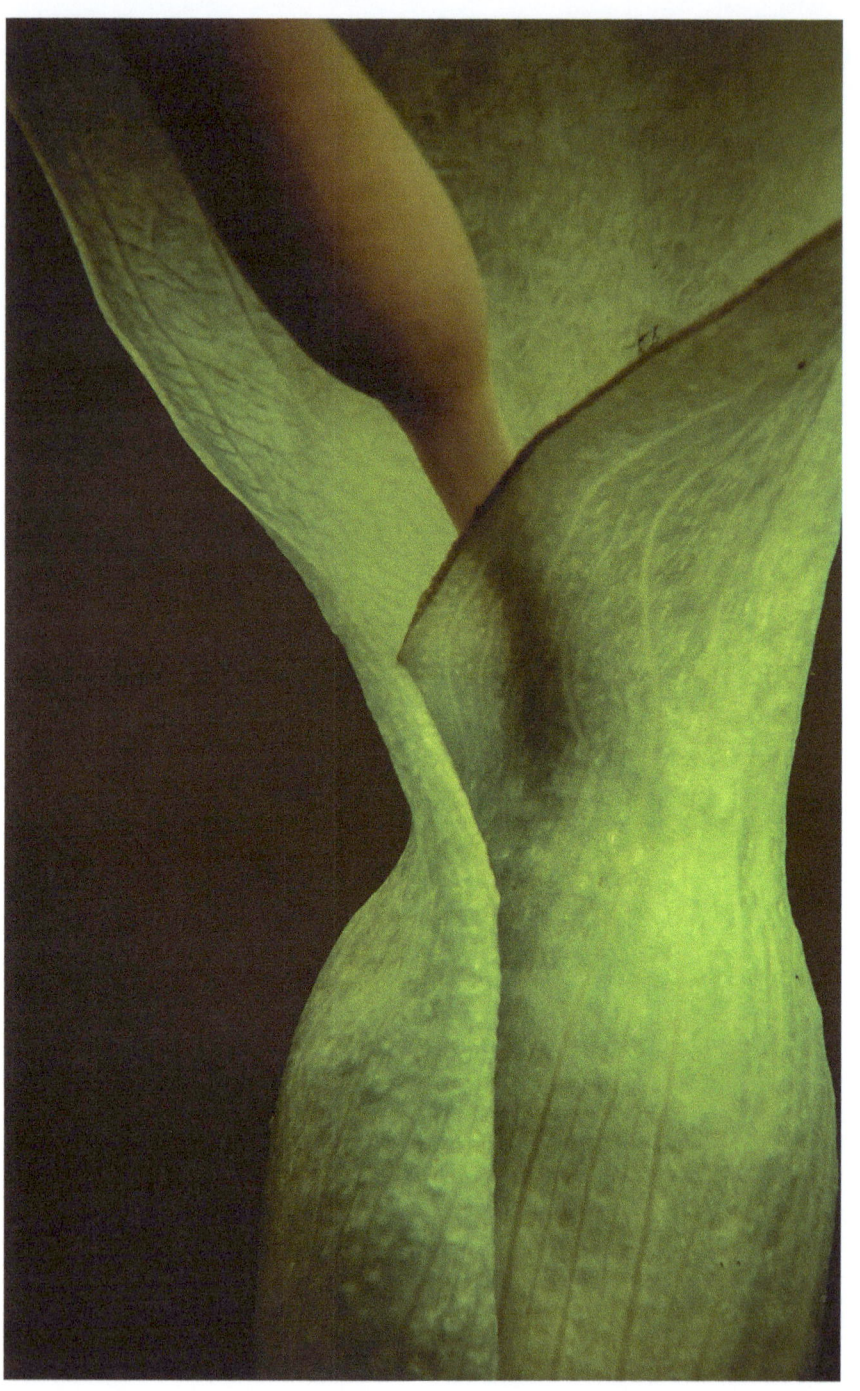

16

A rum is a botanical archetype standing in the shadows beneath the ancient yews. Wrapped within the green cowl we see flashes of a swirling serpentine power. In our art and literature Arum is a symbol for the fiery and potent energies of sexuality, fertility and death. No wonder that this plant continually tugs at our deepest motivations and fears. Arum nags at us, pushes at us, refusing to be overlooked, insisting on being given shape in our words and images and stirring a continual response in our art and literature and mythology.

We begin, as before, in ancient Egypt. Here exist the very earliest artistic depictions of Arum dating from an incredible 3500 years ago, give or take the odd century. In the desert north of Luxor, the Egyptians produced what was effectively one of the earliest known illustrated plant guides, not as a written manuscript but as a collection of stone bas relief carvings within the temple complex of Karnak. Thutmosis III, who ruled Egypt from 1501 to 1447 BCE, was clearly a cultured man who encouraged botanical surveying of the lands he had recently conquered and the results of these surveys were 'published' in some 275 carvings in the appropriately named Court of Flowers at Karnak. Two of these are of Arum and actually depict two different species. Not only is it unusual to find such a modern distinction being made this early in history but these carvings are the earliest pictorial representations of Arum in existence. Given their dual nature of artistic value and botanical plant guide, these stone reliefs are the almost literal bedrock of our survey of Arum-inspired creativity and as perfect a foundation as we could ever hope to have.

After the Egyptian tradition comes the Greek impetus; from the time of Dioscorides onwards through the Middle Ages, Arum continues to be depicted in the ever more garbled illustrations of the ancient herbals. Even when copied for the thousandth time and depicted as a rather abstract hollow tube with a point at the top, Arum is still just about recognisable, even if only because it simply could not be confused with any other plant. This general state of visual literacy didn't really improve until Fuchs published his botanically accurate illustrations in his herbal of 1542, and even this was only possible because he instructed his draughtsmen to draw what was there and expressly forbade them to indulge their artistic fancies. A rather radical approach for the time but the results spoke for themselves and provided some of the most beautiful botanical drawings of plants (and Arum) ever produced, as well as the first usable illustrations in a herbal for over a thousand years. Generally however, it's fair to say that, despite Fuchs' notable exception, the depictions of Arum in herbal literature have more in common with 'art' than illustration, even if this is mostly unintentional.

Away from the practical purposes of the herbals, we move into more purely artistic depictions of Arum. Quite possibly the most beautiful example of this and one where Arum is depicted accurately as well as artistically, is in the last of the so-called 'Unicorn Tapestries'. This series of seven tapestries represents one of the pinnacles of medieval art, each hanging measuring just under 4 x 3 metres, woven in wool and silk with silver and gold threads. Created between 1495 and 1505 in Brussels or the Netherlands, it is still a mystery why and for whom they were made or even whether they were originally intended to be hung together at all. It is known that for centuries they were in the possession of a French family by the name of La Rochefoucauld, who hung them in their family home in Bordeaux. In 1789, amidst the chaos of the French Revolution their chateau was ransacked and the tapestries stolen.

The family returned to their chateau in 1855 and made it known around the locality that they wanted the tapestries back and would pay for their return. No questions asked, presumably. Astonishingly, all of the tapestries were still in the hands of families of local farmers and the Rochefoucaulds regained possession of their wonderful hangings. Even more remarkably, it turned out that these incredible tapestries, at this point some 350 years old, had been used by the local farmers like old rags; for protecting piles of potatoes in barns and wrapping around fruit trees in spring to protect them from frost. For 65 years. Despite such unintended and seemingly abusive activities, the tapestries were still in good condition and generally had not suffered undue harm. The family hung them in their home once more until 1922, when they sold them to John Rockefeller for $1 million. After enjoying them in private for 15 years, he donated the tapestries to their current home.

The tapestries are today displayed in the Metropolitan Museum of Art (in New York) and depict the hunting and killing of a unicorn by a group of noblemen. The sixth tapestry shows the unicorn as dead; killed by the noblemen and brought back to the castle for the lord and lady. In keeping with Arum's symbolic powers, however, the seventh and final tapestry depicts the unicorn as alive and well, seemingly resurrected but now contained by a fence and chained. It is, however, a very low fence and the chain is very loose; the implication being that if it wished to escape, it could do so. Many plants are depicted within the fence with the unicorn and Arum is one of these, very visible, in full fertile glory, beside the resting unicorn.

This particular tapestry (and Arum) also makes an appearance in the film *Harry Potter and the Half-blood Prince*, where it is seen covering the wall outside the 'Room of Requirement' with the character Malthoy standing before it. It is also glimpsed against the walls in various Harry Potter 'house' common rooms. A high-resolution image of the seventh tapestry depicting Arum can be seen on the Metropolitan Museum's website at http://bit.ly/Lg2Zt3.

Contemporary instances of Arum in art include the work of the artist John Nash who, along with his marvellous flowing and sensual landscapes, produced a beautiful woodblock print of the British Arum in 1927, quite likely based on a plant found in the area around his Sussex home, the landscape of which inspired so much of his art. It is tempting also to include Georgia O'Keeffe, who painted such enveloping images of flowers, very often from the Araceae family. But these were the more exotic, tropical or American types of Arum lilies, rather than the British or European varieties of Arum, though the symbolism of her paintings is perfectly in harmony with the nature of the Arum plant.

On that point it should be mentioned that, across the wider family of Araceae, there is a vast wealth of artistic and photographic representation; its members are far too beautiful for this not to be the case. Arum lilies inspire with their pure and serene beauty, while the Titan Arum overawes with its sheer size and grotesque fumery. To provide even a superficial overview of such a wide spread of subjects would take an entire book in itself. Instead, in this book our focus is on the more familiar, the more archetypal form of 'Wild Arum' and principally on the type found here in the UK and northern Europe and, to an extent, in similar climates elsewhere. In this way, we can try to uncover some more of the meaning which this plant carries and tries to convey to us through its nature and virtues.

Arum 'type' plants also make an appearance in architecture from time to time. In Paris there is a building at 33 rue du Champ de Mars designed by the architect Octave Raquin. It is a marvellous example of the Art Nouveau tradition and sweeps itself about in curving floral lines and decorations. So much so that it is known as the Maison des Arums or 'house of lilies'. While the carvings are generally more lily than Arum, there are some instantly recognisable Arum plants above the double doors at the front of the building. A collection of these images can be seen at the flikr set of the user 'dalbera' at bit.ly/10Z1Jby. Arums also appear in the library of the incredible 'House of Scientists' in Lviv, Ukraine.

A more specific example can be seen in a church in southern England. In the early 1900s, members of the then current Arts & Crafts Movement were involved in the design of Four Elms Church in Kent. Inside is an image of Arum in full 'flower', carved into the pews. It is said to have been carved by Evelyn Chambers, a member of the Women's Art Workers Guild. Richard Mabey also relates that there is a carving of Arum in its fruiting berry phase in Westminster Abbey.

The wider Arum family is also responsible for inspiring the creation of a series of beautifully stylish floor lamps by the designer Sandro Santantonio and the lighting company Lucente. The lamps stand 2 metres high with the 'flower' part in fabric and the metal stem containing the lighting elements, one pointing upwards and the other downwards. The lamps embody, with all the skill of Italian design, the inner essence of Arum laid bare and illuminated with inner light. More images can be seen on the website of the designer (under the 'light' menu at www. sandrosantantonio.com) and on the website of the lighting company Lucente.eu at http://bit.ly/WjLgtV. There is also a flikr set at http://bit. ly/OyKoQA.

Outside of the literature of the herbals, Arum has often appeared in literary works such as plays and poetry. The chapter on folklore uncovers some of the references to Arum in the plays by Shakespeare and Lyly that make use of Arum's more esoteric and sexual symbolism, and Arum

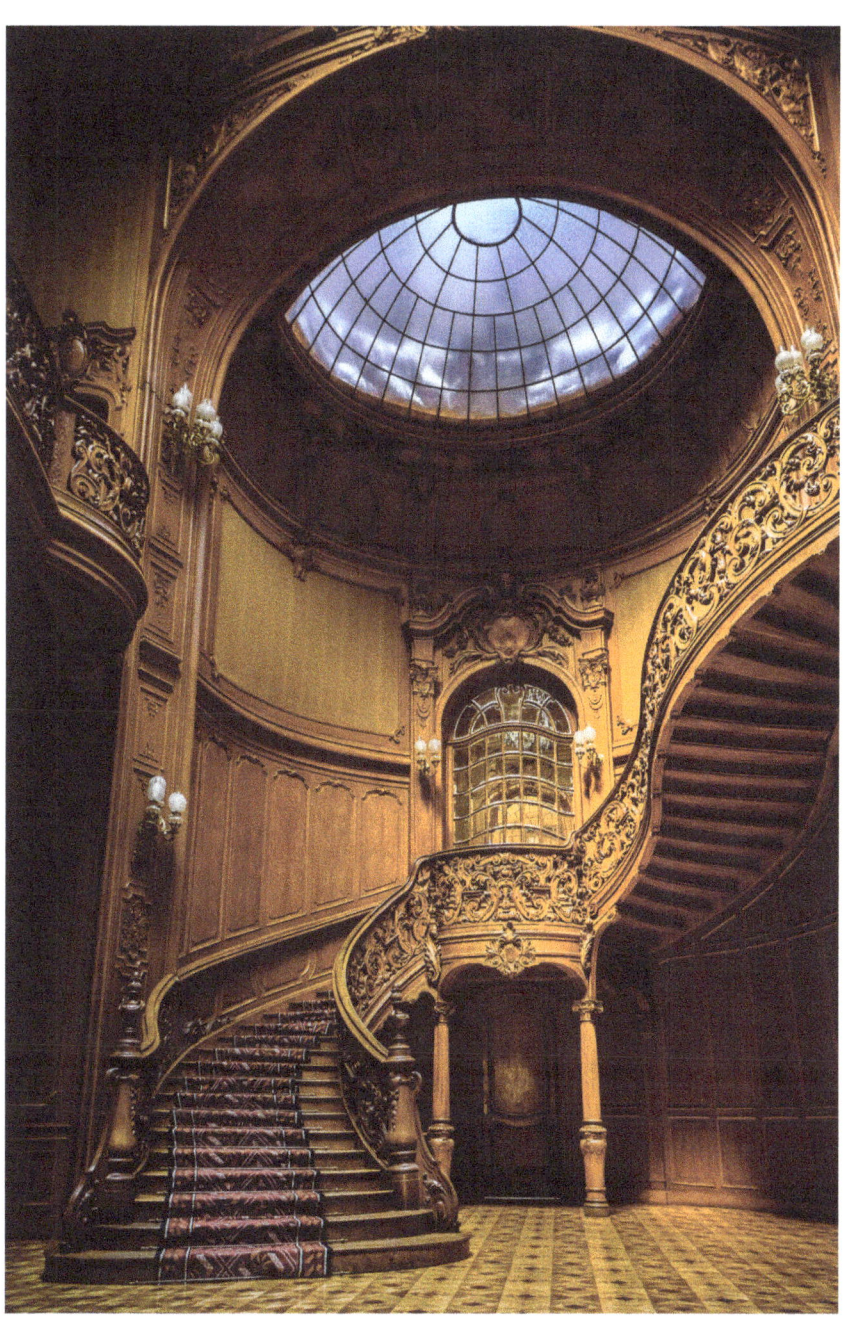

The House of Scientists, Lviv. At the top of the stairs is the library.

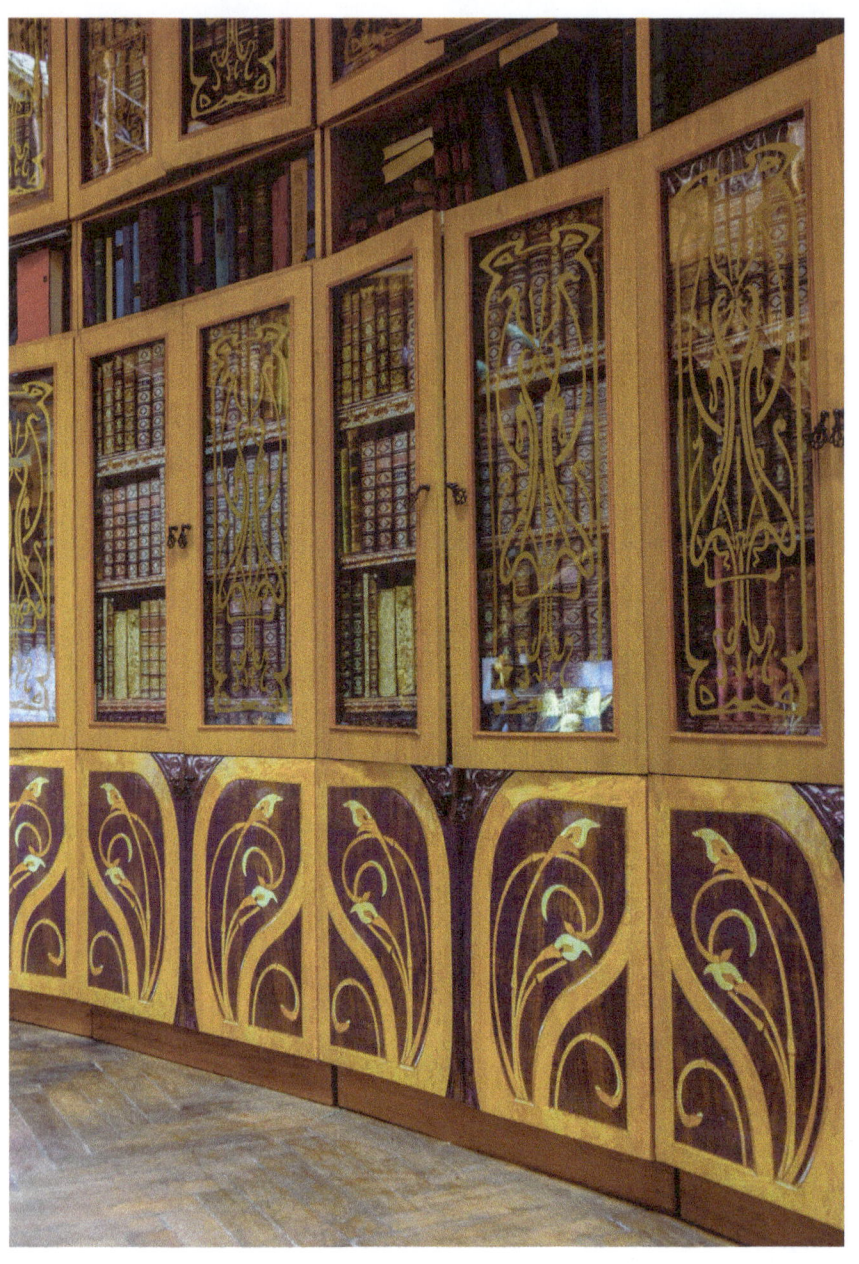

The library room in the House of Scientists, featuring Arum inspired book shelves.

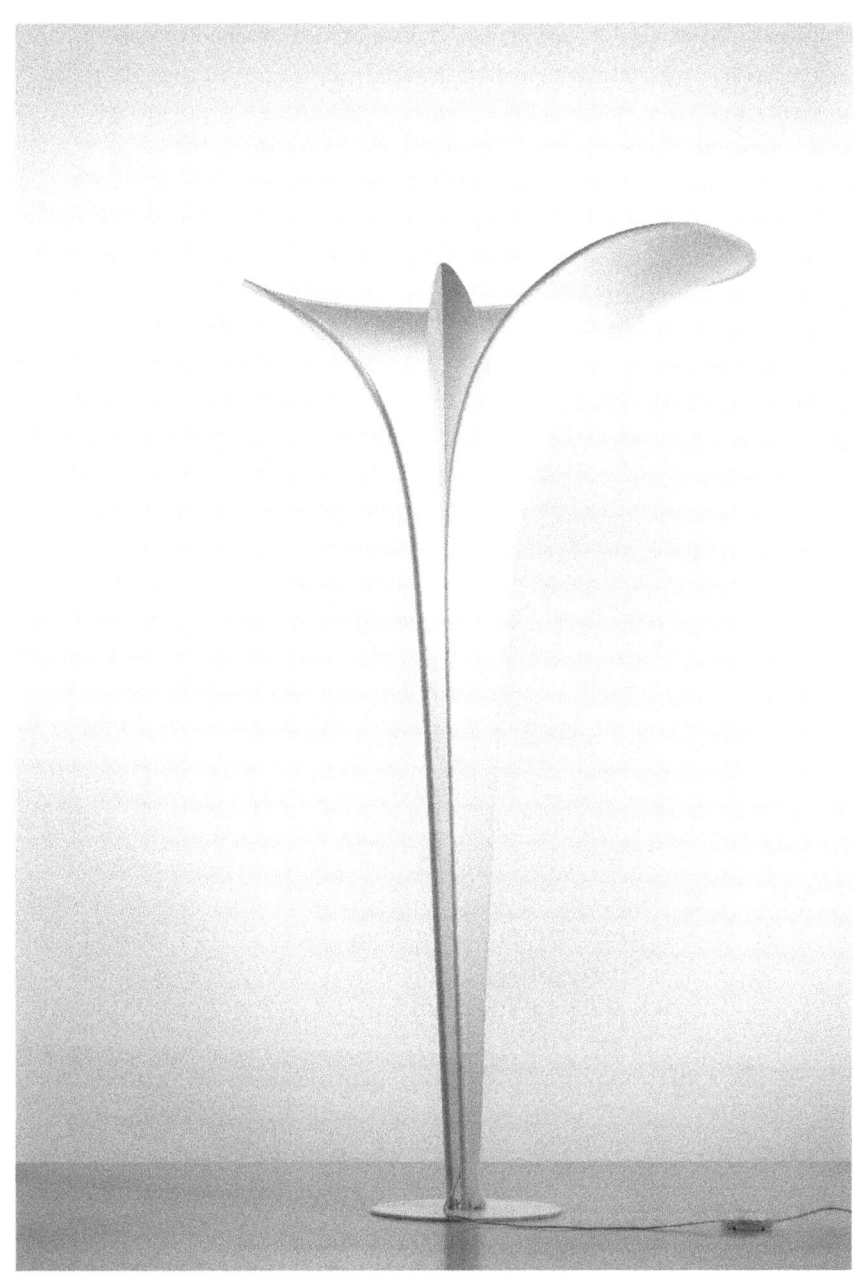

The Arum inspired lamps from Lucente in white.

might also have been mentioned by Thomas Nash in his dismissal of
those lawyers who starch their beards. Shakespeare is also thought to be
describing Arum in his play *Love's Labour Lost*, in the song of spring:
'When daisies pied and violets blue
And lady-smocks all silver-white
And cuckoo-buds of yellow hue
Do paint the meadows with delight.'

Thomas Meehan considered this to be referring to the old
association of the darker-coloured English Arums (the 'Lords') and the
lighter ones (the 'Ladies') in all of the dual-type country names of Arum
and with which Shakespeare was playing in this passage.

Unsurprisingly, Arum also makes itself known in a great many
poems, as a search for 'arum' on poetry sites will attest to. Among the
more notable examples are Anne Barbara Ridler, who mentions Arum in
her poem *The Spring Equinox*, and Dana Ward in her wonderfully titled
Flowers of the Foothills & Mountain Valleys II - Rise of the Demons. The
Scottish poet George MacDonald, who inspired many fantasy writers such
as Tolkien and Lewis, writes about an 'ever dreaming arum' in his poem
Wild Flowers. It is not just historical writers who felt Arum's influence. On
the website www.poetryhunter.com are also a great many contemporary
submissions including many from writers in more tropical countries such
as India, who feel compelled to include Arum in their poetical writings. All
of these and more can be found on www.poetryhunter.com.

One of the most delicate placements of Arum is in Caroline
Southey's poem, *The Primrose*, where she writes:
'An oak's gnarl'd root, to roof the cave
With Gothic fretwork sprung,
Whence jewell'd fern, and arum leaves,
And ivy garlands hung.'

The American poet, Philip Henry Savage (1868–1899) wrote about
the spirits of the earth hiding beneath the Arum leaves during the day,
enjoying the cool air to be found there. The inner nature of Arum is also
stumbled upon by Charles Mair, writing in 1840 in his *Lefroy in the Forest*:
'This is the Arum, which within its root
Folds life and death'

A beautifully evocative poem by the English poet Ted Walker can
be found in his book *Fox on a Barn Door*. Ted Walker was a West Sussex-
based writer and poet notably active from the late 1960s to the mid-

1970, though his literary and creative output in total covered almost five decades. *Fox on a Barn Door* was his first of five critically acclaimed books and contains his short poem *Cuckoo Pint*, in which he describes Arums as 'bright hedgerow tarts ... vulgar and brazen'. It is worth buying his book for this poem alone, but his writing in general is highly recommended.

The finest poet to repeatedly bring Arum to life in his work, however, is undoubtedly the English rural poet John Clare, a fascinating resource about whom can be found at www.johnclare.info. John Clare wrote about many subjects: nature, oppression, mental health, politics and the society around him. After initial success he experienced obscurity and poverty combined with ill health, yet this did not stop him from writing. His life makes for a painful and poignant read. He spent many years in a mental asylum yet produced some of his finest work there. Four of his poems refer to Arum, the first at the beginning of this book, *On the Sight of Spring*, which Anne Pratt also used to preface her book *The Flowering Plants of Great Britain* (1855). In another springtime poem, *The Shepherd's Calendar – April*, he writes:

'To see thy first broad arum leaves
I lovd them from a child.'
Clare also mentions Arum in his lengthy *The Village Minstrel*,
in part 10, where he writes,
'And hooded arum early sprouting up
Ere the white-thorn bud half unfolds to view.'

And now, after having begun with names and travelled through the biography of Arum's life, there remains the final and most subtle aspect to explore: Arum in mythology and folklore.

Further Reading.
See the links provided within the chapter.

CHAPTER 17

THE ARUM OF MYTH & FOLKLORE

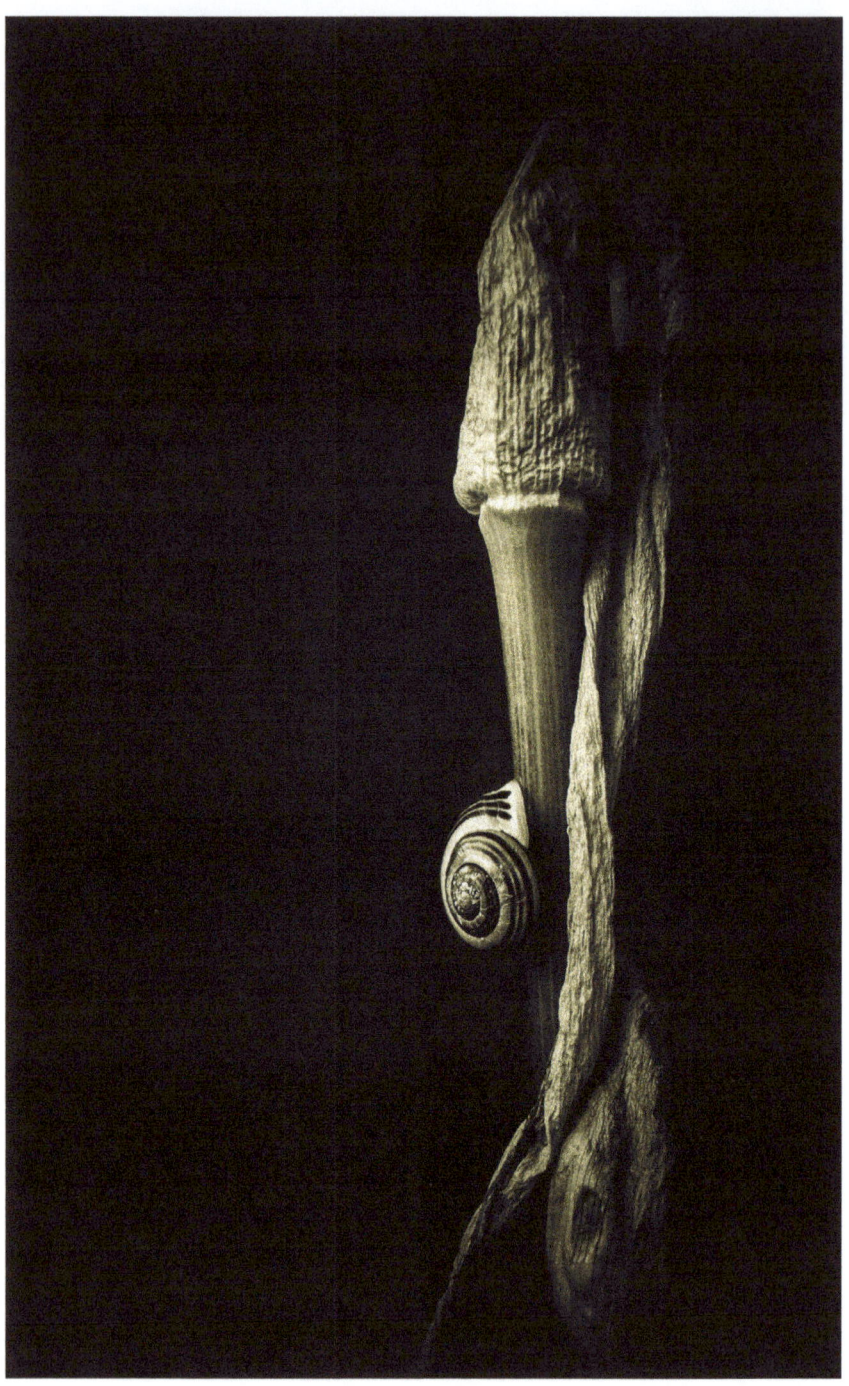

17

Here we are, at the fading edge of the sunlight. Ready to leave this safe and stable world of the known to enter the shifting world of the maybe, the perhaps and the 'let's hope not'. Let's slip beneath the trees into the world of folklore and myth. Because Arum is in there as well, waiting for us. And we can't turn back now. We're almost at the end.

In the shades of our woods we find it, standing like a spirit from the underworld given shape in ours. A seasonal manifestation of something decidedly 'other'. Its strange, hooded form challenging us to explain its presence. Is it friend or is it foe? Or is it merely indifferent; a being busy with its own unguessed-at business?

What symbolic mystery is the Arum portraying in its strange shape? What meaning is it carrying? Our ancestors thought they knew. Sustained in our folklore, Arum brings us face to face with our most chthonic desires and fears: those life-twins of sex and death. From life-creating sexual desire to death-dealing serpents, Arum is there, midwife of the doors to and from this life, encouraging us to create and saving us from death. Sometimes. The shaman plant once again.

Let us keep our wits about us then, but not too tightly, and step through these doors into the shadow space of our collective unconscious, into that oral repository of knowledge, both known and unknown, that is the shared inheritance of our own folk history.

If myth is the explaining of big questions about the world, folklore is the filling out of the detail. For most of our history, our knowledge has been passed down the generations in our folktales, in one great oral relay of never-ceasing conversation. Its origin disappears into our own lost history, for we do not know, really, where we came from or how we gathered our intimate knowledge of plants and their uses. In most mythologies, the gods themselves taught this knowledge to us. One of our lessons was undoubtedly about symbolism.

One of the earliest ways in which our ancestors made sense of the world was through the 'Doctrine of Signatures'. This stems from a belief, ancient beyond history, that this world reflects, symbolically, the world of the gods, or at least, some other, supranatural realm. Everything we need is here, if we have but eyes to see it. For herbalists this means that plants resemble that which they treat. Kidney beans are so called because they clearly resemble kidneys. Walnuts resemble brains. Lungwort resembles a diseased lung. Eyebright is so named because it resembles the brightness of the clear eye. Dioscorides, despite his unusually modern approach, did not think to question this, so self-evident was the truth (and usefulness) of this view. And as we know, everyone following followed Dioscorides.

The very earliest writings convey the usefulness of Arum in treating bites and stings, particularly from serpents. Serpents carry possibly more lore and symbolism upon their long backs than any other creature and embody one of our most instinctive fears. Such instinct is well placed. Even today snakes kill over 100,000 people each year. Those who survive often experience such severe tissue damage where the bite struck that they are left with permanent physical disability and deformity. Snakes can be dangerous if threatened or come upon unexpectedly, no doubt about it, beautiful creatures though they are. One thousand years ago the risk from poisonous snakes was that much greater even in Europe and they were a genuine and frequent danger.

'Like cures like' according to the Doctrine of Signatures, and Arum bites like the most dangerous kind of serpent. Even its leaves are spotted, like the viper, and hooded, like the cobra. The signs are obvious and overwhelming and only a fool ignores them. Thus was Arum prescribed across the ancient world for snakebite. This originally oral knowledge grew in the telling into a prescription that a person need only carry a piece of Arum root to scare away serpents and even burning the plant would drive them away from one's presence. Is Arum useful in treating venom? As far as we know, no. But we don't know very far because no one today has tested Arum for anti-venom properties. But remember that Dioscorides also advised that merely smelling the flowers of Arum could cause an abortion and we have already seen the truth in that. While it is unlikely that carrying a slip of Arum root will send nearby serpents scurrying into the undergrowth, perhaps one day a diligent researcher will discover that Arum does indeed have anti-venomous properties and the Doctrine of Signatures will chalk up yet another notch of success. Until then, this use must remain as folklore. Though Arum does, as we have seen, have eerily similar venomous qualities in its toxins. Perhaps we'd better pick a little bit then, just in case.

If medicinal folklore focuses on Arum's ability to save us from death by snakebite, popular folklore focuses altogether on the opposite end of life's stream: sexuality. The potent physicality of this strange plant, impudently displaying itself in unashamed openness is self-evident. How could Arum be anything else but a powerful aphrodisiac? Dioscorides could not ignore it and counselled that when drunk with wine it stirred up 'vehement desire to sexual intercourse'. That's pretty strong. A 'vehement desire' is one which will let nothing stand in its way and will not relent until fully satisfied. But then, how could a plant with as priapic a form as Arum induce anything less?

The power of Arum to induce and symbolise unabashed sexuality is rampantly evident in the rich fecundity of its local names alluding to this: Dog's Dibble, Gentleman's Finger, Stallions and Mares, Willy Lilly. In this vein it is a plant which simply stands there and exposes itself for all to see. It shouts about sex and intercourse, fucking and ploughing its way through the Anglo-Saxon imagination, its very existence a living symbol of the sex act made real. It invites all who see it to be as open with their sex and sexuality as it is. 'Release your desires!' it cries. 'Rejoice in your springtime couplings and in the life and delight that your carnal celebrations create'. The maypole of plants with a hot-line to Eros and Aphrodite both.

Yet it also symbolises an aspect of the erotic which is far removed from such earthy rutting. For in its physical form it displays very vividly male and female sexuality joined together as one. The 'pintle in the cowl' is as good a symbol as any of the male and female forces each protecting and benefiting the other with their respective qualities and strengths. Though the manner of display is earthy and obvious, the meaning is spiritual and subtle. It is the union of opposites made real while still remaining in and of this world. In the midst of ribald exoteric sexuality, every Arum in every wood, along every hedgerow, openly displays a deeply esoteric message. For those that have eyes to see, as would say the ancient philosophers.

The viagric nature of Arum, and the 'vehement sexual desire' it induces, were well accepted in 15th and 16th century England, and this was reflected in the literature of the time.

John Lyly's play, *Love's Metamorphosis*, published in 1601, concerns the Greek myth of three foresters who fall in love with three nymphs who happen to be daughters of the Goddess Ceres. Arum is mentioned in the very first act, by its name of 'Wake-Robin', but at this stage in the play it is not love for the nymphs which Arum has awakened in the hearts of the foresters, but a love of hunting:

'Niohe: Come, let us make an end, lest Ceres come
and find us slacke in performing that which wee owe.
But soft, some have beene here this morning before us.

Nisa: The amorous foresters, or none; for in the
woods they have eaten so much wake-robin, that they
cannot sleepe for love.

Celia: Alas, poore soules, how ill love sounds in their hips,
who telling a long tale of hunting, thinke they have bewray'd
a sad passion of love!

Niohe: Give them leave to love, since we have libertie
to chuse, for as great sport doe I take in coursing their
tame hearts, as they doe paines in hunting their wilde harts.'

Beware then, those who eat of the Arum, for it is desire itself which it awakens.

Wake-Robin is from the old French word *robinet*, meaning penis. In modern French this means 'tap', and the similarities to the falling spadix after fertilisation are ripe. Wake Pintel is a similar and related name. Arum also appears in Shakespeare, when Ophelia cries 'For bonny sweet Robin is all my joy'. She is gone mad with desire for her lover's 'robinet'. Vehement sexual desire indeed.

A faint echo of this folk knowledge of the 16th century was still alive in the 1930s in rural Dorset (the source of much Arum-related activity and home to the Isle of Portland). Young girls were told never even to touch the plant because they would then fall pregnant if they did so. On one level it is perhaps a sensible warning to scare youngsters into avoiding the irritating crystals in the leaves, while on another it is lingering traces of folk memory – a warning against awakening an unstoppable sexual thirst in otherwise quite 'proper' young Christian girls from Dorset.

Culpepper relates an unusual tradition concerning Arum. He describes a German belief that 'if a young man places the leaves of Arum in his shoes before going to a dance, he will draw to him the prettiest girl in the room. He should recite the following rhyme when placing the plant in his shoe:

'I place you in my shoe, Let all the girls be drawn to you.'

He also says that Arum, astrologically, is under the dominion of Mars, due to its biting and fiery taste. Mars is, of course, the God not just of war but of passion and lust. No wonder he rules over Arum. It's just his thing.

Arum also appears in biblical legends according to Skinner (1911). He relates the tale of the 'spies of Israel' who go into the promised land carrying Aaron's rod which they used to carry back a large bunch of grapes. Upon arriving back, they stuck the rod into the ground whereupon Arum sprouted, symbolising the abundance they had brought back. The arum plant still carries this symbolism of fertility and abundance and Skinner says that farmers use it to gauge a season's coming fetility, predicting the size of their crops by the size of Arum's spadix (or rod). Proving once again that size really does matter.

After love's climax comes the fall and Arum displays the fall of priapism so literally it must make the very Gods themselves blush. Having fulfilled its purpose, the purple-red pintle loses its turgor and begins to fall, to soften and deliquesce, to droop and shrink, to shrivel to a shadow of its former glory. For a while, the arrow-shaped cowl within whose curves it once stood proud, guarding the entrance to the chamber below, bends over protectively, as if still to hold and nurse the dying pintle a while yet, until it too begins to wither as the plant itself now begins to undergo its own metamorphosis.

After life comes death. And Arum displays this with no less directness than it does that of sexuality. The once beautiful and stylish form now turns into an exhibition of monstrous decay and decomposition. No turning a beautiful autumnal colour for the Arum. Instead it becomes its very own grotesquery, drooping, twisting, drying or rotting as it sees fit. Its very aim seems to be to become the opposite of its previous springtime beauty. From enticing maiden to repulsive hag. From Adonic youth to ancient 'death with scythe' made manifest. No wonder the ancients used to rub Arum along the blades of their scythes to keep them sharp. Here is displayed the fleetingness of sexual joy and intercourse, followed by its consequences. Arum is acting out the entire cycle of life as if there is no tomorrow. And perhaps, for Arum, there isn't. We can only watch and learn. For a time, there is little to see.

Arums are sometimes known as Janus plants after the Roman God Janus, who had two faces and was thus a symbol of a threshold guardian, opening access between this world and another. It is notable that folk intuitions have sensed the hidden nature of this plant and linked it to this appropriate symbol.

Unlike other plants, Arum does not have a strong association with witchcraft or sorcery. Only one record comes to light, by Richard Folkard in 1884 who, in his book on plant mythology states:

'The chief strength of poor witches lies in the gathering and boiling of herbs. The most esteemed herbs for their purposes are the Betony-root, Henbane, Mandrake, Deadly Nightshade, Origanum, Antirrhinum, female Phlox, Arum, Red and White Celandine, Millefoil, Horned Poppy, Fern, Adder's-tongue, and ground Ivy. Root of Hemlock, digged in the dark, slips of Yew, slivered in the moon's eclipse, Cypress, Wild Fig, Larch, Broom, and Thorn are also associated with Witches and their necromancy.'

Folklore is less exuberant on Arum's deathly qualities but Christianity, that modern-day religion of death, knows where to find its symbols. The spots on Arum's leaves and the purple-red colour of its spadix are caused, it is said, because Arum grew at the foot of the cross and, where the drops of Christ's blood fell on to the plant, they stayed and gave the plant its colours.

Thomas Meehan, in his book *The Native Flowers and Ferns of the United States* (1878), quotes the following story about this legend from a Mrs Hemans:

'Beneath the cross it grew;
And in the vase-like hollow of the leaf,
Catching from that dread shower of agony
A few mysterious drops, transmitted thus
Unto the groves and hills, their sealing stains
A heritage, for storm or vernal shower

Never to blow away.'

Which is interesting, because Christianity is not just the religion of death. It is also the religion of resurrection. And this is just what Arum does next.

After all the activity and display and commotion of its early life, Arum then promptly goes quiet. The activity has shifted to the invisible underworld. All that is left are tall, green stems topped with tightly packed bunches of small, green berries. Innocuous against the forest greenery, they stand liked stored herbal maces, ready for use. Silent. Waiting.

After the pause comes the last push of activity. In later summer and early autumn the underworld marriage is complete. The Queen of the Dead has lost her virginity in her underworld marriage bed. Arum reflects the event into this world, setting its green maces afire with lip-red berries swaying in celebration. The red of the underworld and the blood of rebirth. No wonder then, in recognition of this fecundity of life, that in Germany, that source of our most famous of folk tales, they say that where Arum flourishes, the spirits of the wood rejoice.

Finally, it remains to mention the supreme Egyptian God of all Creation, whose name of Atum is also translated as Arum. Arum is the Egyptian equivalent of Zeus, Jehovah and all the other ultimate creator deities. In Egyptian mythology, Arum created the universe out of the waters of chaos. His tears of love fell to earth and created man. The scarab beetle, of which there is a large carving at Karnak (where Arum the plant is also depicted), is sacred to the God Arum.

Arum is a god from the early and original Egyptian culture. He later mythologically merged with Ra but, in keeping with Arum's more shadowy nature, Arum is the god of the setting sun who lifts up the dead into the heavens. The shaman plant once again. In the Book of the Dead, it is foretold that Arum, as the original creator God, will at some point exercise the flip-side of that creative power and destroy the entire world.

Interestingly, the God Arum sometimes took the form of a serpent, particularly when feeling creative. Given (the plant) Arum's long association with serpents: as a protection from them, as a remedy for their poison and in its possession of snake venom-like poison, we could justifiably conclude that Arum is really a serpent plant; perhaps symbolically a snake in plant form and the representation in this world of the ultimate creator God, granting us access to the creative and regenerative powers of the serpent, including being a doorway into and out of this world we know.

In a wonderful circle of symbolism, Arum is fertilised by owl midges. Owls are the symbol of Ariadne, the Greek Goddess whose name means 'hand of fire'. Arum's reproductive process is one which revolves around a fiery heat and its name is derived from one linked to the Arabic word for fire and the Hebrew word for serpent. Ariadne is traditionally depicted as holding two serpents in her hands. The swirls of symbolism spiral around and around.... So be kind to the Arum you pass on your woodland walks. It could be the ultimate God of all in disguise.

Further reading.

Skinner, C. M. (1925) Myths and Legends of Flowers, Trees, Fruits and Plants in All Ages and in All Climes.

Folkard, R., (1884). Plant Lore, Legends, and Lyrics. Embracing the Myths, Traditions, Superstitions, and Folk-lore of the Plant Kingdom.

Cristhwaite, H. (2007). A Fire Not Blown: Investigations of Sacral Electrical Roots in Ancient Languages of the Mediterranean Region.

CHAPTER 18
FINAL WORD

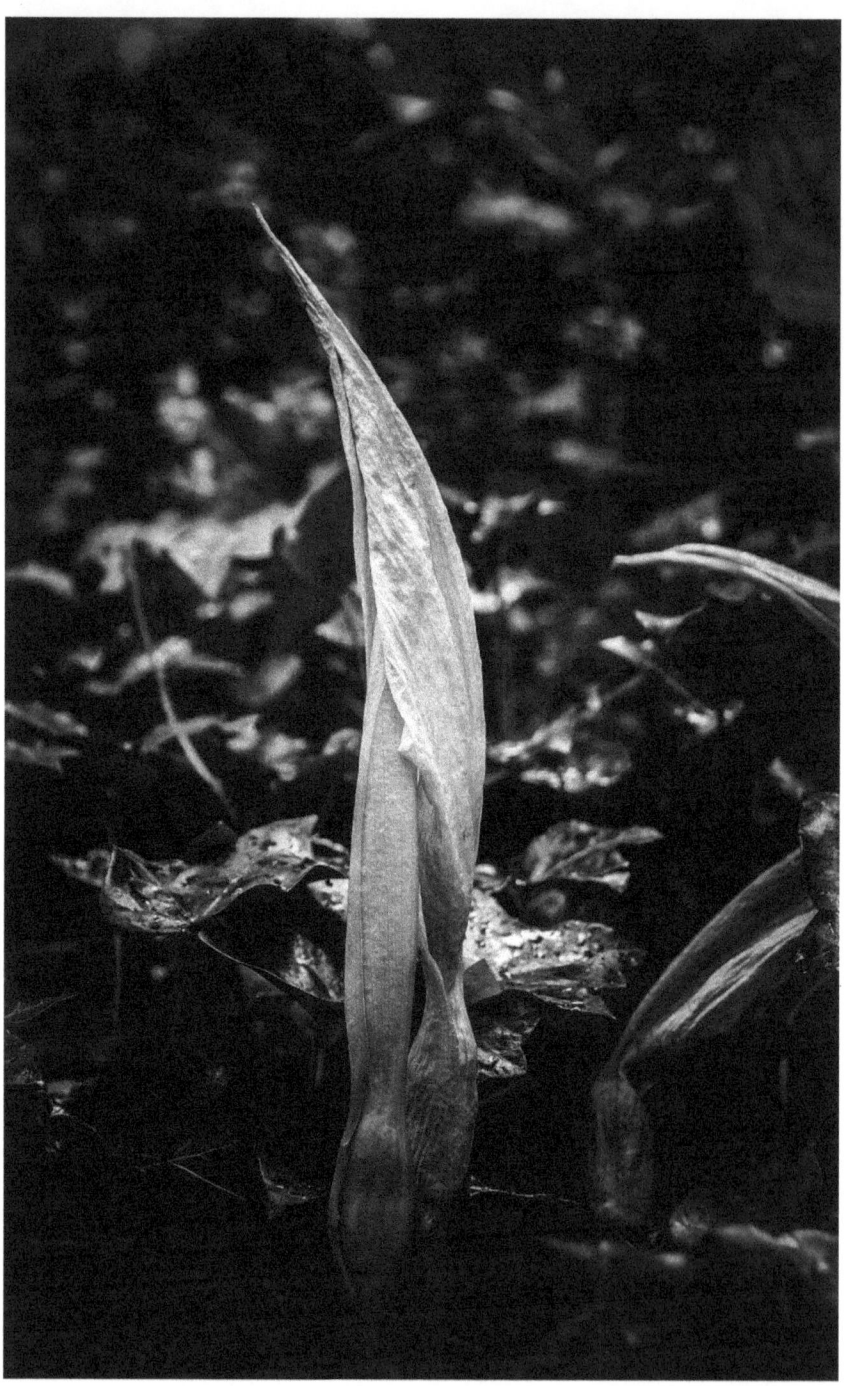

18

We have come to the end of our journey with Arum. And what a journey it's been. It has let us get to know it, just a little, and left us reeling with what we have seen. Fire. Sex. Serpents. Life and Death. The forces of Creation and Destruction. It seems that these are the qualities of Arum. Who knew we had the 'Creator God' incarnate living with us in our local woods? Or that Elizabethan fashion was so tied to the rise and fall of a unique industry on a tiny island on the south coast of England owned by the Crown? Or that the plant we pass on our walks consumes itself in a fiery passion during the night? This small, emerald powerhouse embodying the qualities of fire, serpentine power, sexuality, creation and the mystery of life and death. Our countryside walks will never be the same again.

This book has been a joy to write and a fascinating journey of discovery to research. Even photographing the Arum has been an enjoyable process involving getting dirty, wet and midge bitten in the process. All absolutely worthwhile I hope you will agree. In researching the history of Arum I have tried to go back to original sources as much as possible. This has been an interesting task as many, many authors quote Cecil Prime for a great many facts, with several 'sources' forming a circle of self referencing with Prime's book at the centre. Unfortunately, Prime listed very few of his sources though it does look as though he was accessing original source material for many (if not all) of his statements. I have therefore more often referenced the original source material rather than Cecil Prime's book itself.

All the photography has been carried out in the field, in situ where the plants stood growing. The main locations were Malmesbury Abbey Gardens, Tan Pit woods outside of Bristol, the grounds of Wesley College, Bristol and the Isle of Portland. Adobe Lightroom has been used to produce the final images and the 'blackness' behind many of the images is simply black card placed behind the plant; usually held in place by another arum playing a, literally, supporting role.

For those who might take issue with my very loose use of the term 'arum' I can only plead artistic license and state that I am primarily a photographer and not an Aroid specialist and my main aim with this book has been to express just how interesting and beautiful this plant (and indeed its wider family) is. If I have strayed too far from academic seriousness I humbly ask for your forbearance. With regard to Aroid research, if any specialists note that some crucial fact or discovery has not been included I can only ask for forgiveness. This is a particular difficulty with writing a non-fiction book on a subject which is the focus of as much academic study as arums are. As a member of the public a great deal of research is hidden from view behind online paywalls so that if one is not a member of an academic institution (which I am not) then such papers can be, in effect, unobtainable. Should any errors or omissions be spotted, however small, please do let me know so that they can be corrected. It is as easy as sending me an email via the wildarum website. Many thanks and thank you for reading. Lynden Swift 2013.

Further Reading.

At this point, there can be little left to recommend, other than in a most general sense. The one work which can be recommended on Arum maculatum alone is Cecil Prime's 'Lords & Ladies' (1960). This is the only other book solely on this plant apart from the one you are currently reading and is in many ways this book's inspiration. Highly recommended.

For a more modern overview of research on this plant, there is no better source than Peter Boyce's 'The Genus Arum. A Kew Magazine Monograph' (1993). It provides a thorough and easy to read summary of this plant, its wider family and the current (as at publication) understanding of this plant family's many fascinating aspects.

The 2004 paper 'The History & Current Status of Systematic Research with Araceae' (http://bit.ly/A7EmyW), paints the clearest picture of any publication of the vast amount of research being carried out on this plant by individuals and teams from all over the world. It leads the reader on a journey from the earliest historical botanists right up to contemporary researchers recreating the boundary of our knowledge of this group of plants. Essential reading.

APPENDICES
THE TIMELINES

The following timelines provide an overview of the most important or interesting events relating to Arum. They don't attempt to exhaustively list every single event which happened but, instead, present extra information on specific events that do not fit comfortably into the main chapters but do help to provide a better overview of the subject.

The first gives an overall timeline of Arum related events generally. The second and third provide more detailed information on the medicinal and culinary related events respectively.

ARUM TIMELINE

Before Common Era (BCE)

1500 Thutmose III orders carvings to be made at the Karnak temple, featuring two species of Arum.

350 Diocles of Carystos said to have written his herbal.

380–327 Theophrastus, pupil of Aristotle (and both pupils of Plato), writes his *Enquiry into Plants*.

120–63 Crateus' lost herbal believed to have been written.

Common Era (CE)

65–75 Dioscorides' *De Materia Medica* produced.

77 Pliny's encylopaedia. Not a herbal as such but a collection of anything and everything that was known at the time.

400 Herbarium of Apuleus Platonicus.

512 Copy of *De Materia Medica* made for Juliana Anicia.

900 Bald's Anglo-Saxon *Leechbook*. A rare example of a tradition of healing and herbal knowledge in Europe not derived from Greek/Arabic sources.

974 Arum lights up the path along which St Withburga's body was taken after being snatched from its resting place at Dereham church in Norfolk.

1166 *Circa Instans* by Mattheau Platearius. Mostly a translation of Dioscorides' *De Materia Medica*. It was a prime influence on the text of the *Grand Herbier of Paris* (1520) which was itself later translated and printed into English as the *Grete Herball* in 1526.

1250 Physicians of Myddfai. The family tradition from Wales dating back to oral history, written down for the first time.

1256 Albertus Magnus – *De Vegetabilis*. Less a herbal than an original work of early botanical science, though he does mention Arum, stating that it gives protection against all kinds of serpents.

1300 *Tractatus de Herbis* published.

c. 1440 The invention of the printing press.

1481 Herbarium of Apuleius Platonicus.

1495–1505 Unicorn Tapestries produced. Arum depicted in the sixth tapestry.

c. 1500 *Arum maculatum* given its first official name of *Arum officinarum* by

Matthaeus Lobelius.

1525 Bankes' Herbal, the first herbal to be printed in England. Based on an otherwise unknown medieval source manuscript. There were many pirated versions that gave no credit to their original source.

1526 The *Grete Herball*. One of the best-known English herbals, largely a copy of *Circa Instans* and the French *Le Grand Herbier*.

1536 *De Natura Stirpium* by Jean Ruel.

1542 *The Great Herbal* of Leonhart Fuchs.

1551 *A New Herball, wherein are Conteyneed the Names of Herbes, Part 1*, by William Turner.

1554 Rembert Dodoens' *Cruydeboeck*. Translated and reissued firstly by Henry Lyte and then by Gerard.

1558 The Elizabethan era said to have commenced.

Arum maculatum first used as an name by Tabernaemontanus.

Ruffs become part of high fashion, with Arum as the secret ingredient.

1562 *A New Herball, wherein are Conteyneed the Names of Herbes, Part 2*, by William Turner.

1578 *A Newie Herbal*, by Henry Lyte. A translation of the *Cruydeboeck* of Rembert Dodoens.

1583 Philip Stubbes publishes his diatribe against Arum-stiffened ruffs (and other annoyances).

1590 Tabernaemontanus' *Neuwe Kreuterbuch*.

1597 Gerard's Herbal; another translation of Dodoens' work.

1601 John Lyly publishes *Love's Metamorphosis*, describing the amorous effect of 'wake robin'.

1603 The end of the Elizabethan era.

1640 John Parkinson's *Theatrum Botanicum*. Gives possibly the earliest recipes for using Arum purely as a food ingredient; in this case to make an endive and Arum salad and as a meat seasoning.

1651 Culpepper's *Complete Herbal*. Possibly the last of the old medieval-style herbals. From now on books begin to have more in common with modern botanical works.

1657 Coles' *Adam in Eden*.

1658 Topsell's *The History of Four-Footed Beasts and Serpents*.

1686 John Ray expresses doubts about the medicinal value of Arum.

1710 William Salmon publishes *The English Herbal*.

1753 Linnaeus formally names *Arum maculatum* as *Arum maculatum*.
1756 Lobelius publishes his *Stripium Historia*.
1756 John Hill's British herbal.
1796 Mrs Gibbs, of the Isle of Portland, wins the Royal Society competition for finding a new non-food starch source in the form of Arum. A new industry commences.
1798 Bulliard, in *Historie des Plantes Veneneuses et Suspects de la France*, reports the story of the three children of a woodman in France becoming ill with Arum poisoning. Only one survives.
1805 Newton's *A Complete Herbal*.
1838 The last recorded mention of Arum starch production on the Isle of Portland.
1861 *British Medical Journal* reports on three cases of children poisoned by eating Arum.
1870 Anne Pratt records a contemporary use of Arum for flour making in Ireland.
1881 Dr Martindale cures neuralgia using native Arum, thus reproducing the existing cure that used an expensive tropical variant.
1884 *Plant Lore, Legends, and Lyrics. Embracing the Myths, Traditions, Superstitions, and Folk-lore of the Plant Kingdom*, by Richard Folkard (1884).
1885 *Hardwickes Science Gossip* publishes a lengthy article about Arum.
1889 Thistelton Dyer publishes *The Folklore of Plants*.
1897 Thomas Fernie's *Herbal Simples*.
1917 Arum carved in St Paul's Church, Four Elms, Kent by Evelyn Chambers of the Art Workers' Guild.
1927 John Nash produces a woodcut of local Arums.
1931 Grieve's *A Modern Herbal*.
1960 Prime publishes *Lords & Ladies*.
1987 Study on Yugoslavian bears confirms the ancient tales of bears eating Arum after waking from hibernation.
2005–2007 Marcus Harrison conducts experiments on using Arum flour in recipes.
2008-2009 Fergus Drennan carries out his 'wild food for a year' experiment, involving the preparation of Arum flour and its use in various recipes.
2009 www.sarahmelamed.com/2009/02/wild-poisonous-plants-for-dinner/

describes the contemporary use of Arum in Kurdish culture. Arum-inspired lamps designed by Sandro Santantonio and made by Lucente in Italy.

A MEDICINAL TIMELINE

F or a comprehensive timeline of all herbals, a highly recommended source is *The Old English Herbals* by Eleanor Sinclair Rohde (1922).

65-75 CE: Dioscorides: *De Materia Medica.*
　　Recommends the juice of the seeds for earache and for adding to a drink to aid abortion, the root for coughs and the leaves to eat and for wrapping cheese to preserve it. Dioscorides also states that Arum is an aphrodisiac and will excite a vehement desire when drunk with wine. Dioscorides also states that rubbing the root, specifically, on one's hands, will protect one from being bitten by snakes.

77 CE: Pliny's Encylopaedia.
　　Pliny quotes a number of different sources and so includes a large number of uses for Arum, many of which are not repeated elsewhere. The main section is in Book XXIV, Chapter 92, entitled *The Aon: Thirteen Remedies.* Pliny mentions here that Arum can be boiled in milk and used for cloudiness of the eyes, internal ulcerations and inflammation of the tonsils. It is recommended here for freckles, foreshadowing its use two thousand years later by the French as a skin cosmetic. Pliny quotes other sources as stating that Arum is good for difficulty in breathing and general conditions of the lungs and coughs, and restates its use as a facilitator of birth delivery for all animals. He also reaffirms the belief in Arum's efficacy against serpents and snake bite.
　　For more information, see the online version of Pliny's works: http://bit.ly/Yh9Aud

900 CE: *Bald's Leechbook.*
The earliest known written medical lore and the last of the Anglo-Saxon tradition in England:
　　'If a strong potion lodge in a man, and will not come away, take the netlierward part of celandine, and leaves of libcorn

or arod, boil in ale, add butter and salt, give to drink a cup
full of it warm.'
1250: The Physicians of Myddfai.
The family tradition for the first time written down. It recommends
Arum for cancerous growths, by boiling the root in wine and drinking the
decoction over three days.

1526: The *Grete Herball.*
The *Grete Herball* is a fascinating tendril of ancient herbal
knowledge traditions reaching into 16th century England.
Published initially in 1526 by Peter Treveris, it is a translation of a
much earlier French herbal, *Livre des Simple Medicines*, which was already
over one hundred years old when Peter Treveris translated it into English.
The French herbal was itself a translation of an even earlier work
from the 12th century, *Circa Instans*, a work from the Salerno tradition
combining Western (i.e. Greek)-based herbal lore with incoming Arabic
herbal knowledge.
Circa Instans was written by an Italian medical practitioner,
Matthaeus Platearius and is based largely upon a work called *De Gradibus
Simplicium.* This work was written by a Tunisian doctor known as
Constantine the African.
In the medical hotbed of Salerno in Italy, he was a medical
professor who was responsible for translating many works of Arabic origin
into Latin, which had a huge influence on subsequent medical practice
and literature. Constantine's *De Gradibus Simplicium* is an example of this,
being itself a translation of a 10th century Arabic work, *de Gradibus* by a
Tunisia-based Muslim physician known as Ibn Al Jazzar. The stream of
his ideas, through Constantine, eventually finds its way into 16th century
England.
Based predominantly on Dioscorides' great work. around half of
the chapters of the French work are taken from this earlier herbal. The
English translation was the first illustrated plant book to be published in
English. Its description of Arum is in wonderfully archaic English:
'It is called De Iaro. Cucko'we Pyntell, Arus. Calfs foot. It is
hot and dry in the degree. It is also named aron. Some call it
priestes hode. For it hath as it were a cape and a tug in it lyke
a serpentyne of dragos, but serpentyne is longer. It groweth

in moyst places and dry on hyis, under hedges and may be gathered in winter and somer. It hath greate vertue in the leaves, but more in its rote, but yet it hath more vertu in the knottes that be about the rote.'

The medical uses for Arum in the *Grete Herball* mostly mirror earlier works such as those of Dioscorides, but two are worth repeating.

For emozroydes [haemorrhoids]:

'For emozrodes or pyles all evil of the fundiment, seeth this herbe and bathe y Paciet in the same to y navel, Dr bynde the herbes hote in a clothe and let him sit thereon.'

To bring on the menses:

'To cause menstrues to flewe out the iuyce of this herbe into the conduyte with an instrument yzo pze for it oz medle it with the medecine called benet and than bled, oz with cotton wette therein and to minisstered.'

1542: Leonhart Fuchs' Great Herbal.

Fuchs describes the astringent properties of Arum and its efficacy in dissolving sores and growths.

1551: William Turner: *A New Herball, wherein are Conteyneed the Names of Herbes, Part 1.*

'Arum is called in greke aron, in english Cuckopintell, Wake robin or Rampe, in duche Psaffen bynde, in frenche, Vidchaen, the poticarie calleth it Pes vituli, serpentaria minor, luph minus, groweth in euery hedge almost in englande aboute townes in the sprynge of the yere. Some wryte that it is but hote and drie in the fyrst degree, howe be it our aron is hote in the thirde degree.

Dioscorides semeth by hys wryting to thew, that where as he was borne Aro was not so sharpe as it is with us. Galene also writeth, that aron is hote in the fyrst degree. But it that groweth this us is hote in the thyrd degree at the leste. Wherefore some peradventure wyll say, that thys our aron is not it, that Dioscorides and Galen wrote of. But Galene in these wordes folowyng which are wrytte in the second boke: wytnesseth the there are 2 sortes of aro: one gentle another biting.

In certayne regyons after a manner it groweth more bytyng and shapre, to much, that it is allmost as hote as Dragons is and that the fyst water must be eaten out and the roote soden agayne in the second Thys here growynge in Cyrene

is dyfferyng from it, of our countre, for it that is wyth us in
Asia for a great part is shaper then it that growrth in Cyrene.
For the roote purgeth all the inware partes, makyng thyn
and brekyng toughe and gros humours and it is a speciall
good medicine agaynst the almost incurable sore called
cocoethe' It purgeth and scoureth awaye myghtely both
other thyngs that nede scowryng and also the frekelles with
vinegar. The leves also hauyng lyke qualities are good for
freshe woundes and grenesores and the les dry they are the
bytter do they joyne togyther and close up woundes.
For those thynges that are dry, are hoter then that they
can be convenient for woundes. Sum ther be of that beleve
that they thynk if chese be covered with dragon leves that
they preserve it from corruption b the reason of theu dry
complexion. The fruyte is myghtier then the roote and the
leaves. The juice scoureth away the disease of the eyes.'

1578: Henry Lyte's *A Newie Herbal* (a translation of the *Cruydeboeck* of
Rembert Dodoens).

Lyte doesn't say a great deal about Arum because he's already
discussed it under Dragonwort, so Arum as we know it is like a subspecies
in his book. From a manuscript in the British Library, he describes it thus:
'The roots of these herbs either boyled or rotten and
mingled with honey and afterwards licked, is good for them
that cannot fetch their breath. And for those that are vexed
with dangerous coughs and catarrhes, that is to say, the
falling down of humours from the braine to the breath and
against convulsions and cramps, for they divide, ripe and
consume all grosse and tough humours and they cleanse all
the inwards parts.'

All of which looks like a direct copy from William Salmon's book
that came out about eighty years earlier.

Lyte recommends the leaves for eating, after being boiled a
number of times to remove the bitterness, and quotes Dioscorides'
recommendations for Arum's power over serpents and for curing earache
with the juice of the seeds.

1597: Gerard's Herbal (another translation of Dodoens' work).
'The Vertues.
Beares after they have lien in their dens forty daies without
any manner of sustenance, but what they get with licking and

sucking their owne feet, doe as soone as they come forth eat
the herbe Cuckow-pint, through the windie nature thereof
the hungry gut is opened and madde fit once againe to
receive sustenance: for by abstaining from food for so long
a time it is quite shut up, as *Aristotle, Aelianus, Plutarch, Pliny*
and others do write.'

1640: John Parkinson's *Theatrum Botanicum*.
    Parkinson in 1640 views the spotted- and unspotted-leaved Arum as
different species and prefers the spotted kind for medical usage. He has a
chapter devoted to 'Venemous, slepy and hurtful plants and their counter
poysons', which of course if where Arum makes its appearance.

1651: Culpepper's *Complete Herbal*.
    Possibly the last of the 'old herbals' - from now on books begin to
have more in common with modern botanical works. Below is extracted
from a copy in the British Library.
    'Govemment and Virtues. --It is under the dominion of
Mars. Tragus reporteth that a dram weight, or more if need
be, of the spotted wake-robin either fresh and green, or
dried, being beaten and taken, is a present and sure remedy
for poison and the plague.
    The juice of the herb taken to the quantity of a spoonful
hath the same effect; but if there be a little vinegar added
thereto, as well as to the root aforesaid, it somewhat allayeth
the sharp biting taste thereof upon the tongue.
    The green leaves bruised and laid upon any boil or plague-
sore, doth wonderfully help to draw forth the poison. A
dram of the powder of the dried root taken with twice so
much sugar in the form of a licking electuary, or the green
root, doth wonderfully help those that are pursy and short-
winded, and also those that have a cough ; it breaketh,
digestet. and riddeth away phlegm from the stomach, chest,
and lungs : the milk wherein the root hath been boiled is
effectual also for the same purpose.
    The said powder taken in wine or other drink, or the juice
of the berries, or the powder of them, or the wine wherin
they have been boiled provoketh urine, and bringeth down
women's courses, and purgeth them effectually after child-
bearing, to bring away the after-birth. Taken with sheep's
milk it healeth the inward ulcers of the bowels : the distilled

water thereof is effectual to all the purposes aforesaid. A spoonful taken at a time healeth the itch : and an ounce or more taken at a time for some days together doth help the rupture.

The leaves either green or dry, or the juice of them, doth cleanse all manner of rotten and filthy ulcers, in what part of the body soever; and healeth the stinging sores in the nose, called polypus.

The water wherein the root hath been boiled, dropped into the eyes, cleanseth them from any film or skin cloud or mist, which begin to hinder the sight, and helpeth the watering and redness of them, or when by some chance they become black and blue.

The root mixed with bean flour and applied to the throat or laws that are inflamed, helpeth them. The juice of the berries boiled in oil of roses, or beaten into powder mixed with the oil, and dropped into the ears, easeth pains in them. The berries or roots beaten with hot ox-dung, and applied, easeth the pains of the gout.

The leaves and roots boiled in wine with a little oil and applied to the piles, or the falling down of the fundament, easeth them, and so doth sitting over the hot fumes thereof. The fresh roots bruised and distilled with a little milk, yieldeth a most sovereign water to cleanse the skin from scurf, freckles, spots, or blemishes, whatso-ever therein. Authors have left large commendations of this herb yoa see, but for my part, I have neither spoken.'

1657: Coles' *Adam in Eden, the History of Plants.*

Coles introduces the plant and says it is the same plant which others call *Dracontea minor* or *Serpentaria Mi*nor:

'The Signature and Vertues.

The leaves of Wake-Robin, either green or dry, or the Juyce of them, doth cleanse all manner of rotten and filthy ulcers, in what part of the body soever, and helpeth the stinking sores in the Nose called Polypus.

The water wherein the Roots hath been boyled, dropped into the Eyes, cleanseth them from any film or skin, Clouds or Mists, which begin to himder the sight, and helpeth the rednesse or watering of them, or when by some chance, they become black and blew.

The Juyce of the Berries boyled in Oyle of Roses, or beaten

into Powder, andmixed with the Oyl, and dropped into the Ears, easeth pains in them.

The Root mixed with Bean-flowers, and applied to the Throat or Jawes that are inflamed, helpeth them and the Roots or Berries beaten with hot Oxe Dung and applyed, easeth the pains of the Gout.

Tragus reporteth, that a dram or more, of need be, of the spotted Wake Robin, either green or dried, being beaten, and taken, is a most present and sure Remedy for Poyson and the Plague.

The Juyce of the Herb taken to the quantity of a spoonful, hath the same effect, to which if ther be a littel Vinegar added, as also to the Root aforesaid, it somewhat allayeth the sharp biting raft thereof upon the Tongue.

The green Leaves bruised and layd upon any Boyl or Plague sore, doth wonderfully help to draw forth the poyson.

A dram of the Powder of the dryed Root, taken with twice so much Sugar, in the form of a licking Electuary, or the green Root, doth wonderfully help those that are purse and short winded, as also those that have the Cough - it breakth , digesteth and riddeth away Flegm from the Stomack, Chest, and Lungs. The milk wherein the Root hath been boyled, is effectual also for the same purpose.

The said Powder taken in Wine, or other drink, or the Juyce of the Berries, or the Powder of them, or the Wine wherein they have been boyled, provoketh Urine, and bringeth down Womens Courses, and purgeth them effectually after Child-bearing, to bring away the after-birth, and being taken with Sheeps milk, it healeth the inward Ulcers or the Bowels.

The leaves and Roots also boyled in Wine wirth a little Oyl, and applied to the Piles, or falling down of the Fundament, easeth them and so doth the sitting over the hot fumes thereof.

The fresh Roots bruised and diluted with a little milk, yieldeth a most severaign water to cleanse the skin from skurf, freckles, spots or blemishes whatsoever therein.

This plant should be Venerous by its Signature.'

1710: William Salmon's *English Herbal.*
States that Arum's virtues are exactly the same as that of the Dragonworts:
'The Prepared Root.
It is prepared by boiling it till it is soft and all the acrimony

pas'd off. It then nourishes and is good for food, it is also good to expel thick and clammy humours from the brest and lungs. They also restore in consumptions.

The liquid juice of the leaves or root.

It is said to remove the pin and Web, as also spots and pearls in the eyes, being put into collyriums or medicines which are made for the eyes. Dioscorides says that the juice being dropped into the eyes cleanses them and hels dimmness of sight. The same mixed with oil olive and dropped into the ears eases their pain.

The essence of the same.

It has all the virtues of the liquid juice, besides which, being taken inwardly, to one spoonful at a time in distilled water or in white wine, and repeated as oft as need requires, it powerfully provokes the terms in women.

The powder of the root.

If it is made into an electuary with honey, it is good for such as are troubled with vehement coughs catarrhs convulsions, cramps etc., for it incides, absterges and consumes fros, tough and tartarous humors and cleanses all the inward parts. Dose of the pouder from half a dram to a dram. Outwardly applied, it cleanse all fretting and malign ulcers, which are difficult to be healed. It also removes all scorbutick breaking out in anuy part of the body.

The lohoc[1] of the root.

It is made of the root prepared y boiling by beating it in a mortar with twice its weight of honey. It is an excellent things against coughs, salt catarrhs and defluxions of think rheum.

The cataplasm of the fresh and green leaves.

Being applied, it is good for ulcers and green wounds and heals them aftre an admirable manner. Being dry, they are more sharp or biting and not so fit for vulneraries.

The fruit of berries.

They are of a greater power than either leaves or root. And therefore are said to cure virulent and malign running sores and to eat away that cancerous excretion in the nostrils called polypus. They are also good to be laid to cankets and such like fretting, eating and consuming ulcers.

The spiritous tincture of the root.

---

[1] *A Lohoc is a thick medicine which is meant to be held in the mouth to slowly melt, rather than swallowed whole at once.*

Given to one dram or two in any proper vehicle morning and evening it opens obstructions of the womb, and provokes the courses. It eases pains of the stomach and bowles proceeding from wind or from cold, flimy and tartarous humours, prevails against the colick and warms and comforts all the inward parts and is an excellent thing against poyson, plague, spotted fever or any other malign distemper.

The acid tincture.
It is a famous stomach tick, takes away nauseousness and vomiting, warms and comforts a cold stomach, causes a good appetite and digestion and strikes at the root of all poysons vegetable or animal. Especially it refits the biting of vipers and the malignity of mad dogs. It is also a singular thing against malign fevers, spotted fever and the plague or pestilence, by overturning the very fountains of the infectious miasmata. Dose from thirty, forty or eighty drops in the distilled water or any other specifick vehicle.

The oily tincture.
It is good against cramps, convulstions, numnes, palfies, rheumatick pains and aches proceeding from cold and moisture, or in a cold and mosit habit for the body, it is to be anointed upon the parts affected morning and evening and to be well rubbed in.

The saline tincture.
It is good against blackness, greenness, yellowness of the skin, and to take away tanning, sun burning, scurff, morphew , leoprosy, scabbiness, freckles, lentils and other the like deformities of the cuticula or scarrff skinn.

The distilled water.
It is used as a vehicle to convey many of the aforegoing preparations in: it is also a cosmetick, for the beautifying the skin, or to mix other cosmeticks with for that purpose. It prevails also against the pestilence or any other malign and pestilential fever. As also the poyson of serpents or mad dogs, being drunk warm with a dram or two of mithridate or other like antidote with it.'

1884: Plant lore, legends and lyrics, by Richard Folkard.
'A drachm weight of the spotted Wake Robin, either fresh or dry, was formerly considered as a sure remedy for poison and the plague. The juice of the herb swallowed, to the quantity of a spoonful, had the same effect.

Beaten up with ox dung, the berries or roots were believed to
ease the pains of gout. Arum is under the dominion of Mars.'

1885: *Hardwicke Science Gossip* (an early popular science magazine issued
between 1865 and 1893).

'In a certain dictionary published in London by Thomas
Green in the year 1832, and known as *The Universal Herbal
or Botanical, Medical and Agricultural Dictionary* (2 vols), we
learn much. Mr. Green first informs his readers that if they
have been rash enough to taste the 'root', an antidote will be
found either in milk, butter, or oil. Writing still of the 'roots',
he goes on to say:
When dried they become farinaceous[1] and insipid, in
which case they might be used for food in case of necessity;
and by boiling or baking would probably afford a mild and
wholesome nourishment as well as those sorts which are
natives of hot climates.
In being reduced to powder it loses much of its acrimony;
and there is reason to suppose that the compound powder
which takes its name from the plant, owes its virtues chiefly
to the other ingredients. The *pulvis ari compositus*, or powder
composed of arum, is therefore discarded from the London
dispensatory, and, instead of it, a conserve is inserted, made
by beating half a pound of fresh root with a pound and a half
of fine sugar. In the medicine recommended by Sydenham
against rheumatisms, the acrid anti-scorbutic herbs are
largely joined with it.

There then follows a more modern recipe for preparing and using
Arum, along with the result of a personal experiment in doing so. Some of
the ingredients may be a little hard to come by nowadays ...
Dr. Lewis orders the fresh root to be beaten with a little
testaceous powder[2], and mixed with an equal quantity of
gum arabic and three or four times as much conserve, and
thus to be made up into an electuary[3], or else to be rubbed

---

[1] *Meaning that they become essentially starch and little else.*

[2] *A red brick-coloured medicinal powder prepared from the shells of animals.*

[3] *A medicinal substance mixed with honey or another sweet substance, from
the Greek word meaning 'lick up'.*

with a thick mucilage of gum arabic and spermaceti[1], adding
any watery liquor and a little syrup to form an emulsion;
two parts of the root, two of gum, and one of spermaceti. In
this form, he has given the fresh root from ten grains[2], to
upwards of a scruple[3], three or four times a day.
It generally occasioned a sensation of slight warmth, first
about the stomach, and afterwards in the remoter parts[4]; it
manifestly promoted perspiration, and frequently produced
a plentiful sweat. Several obstinate rheumatic pains were
removed by this medium, which he therefore recommends
to further trial. Chewed in the mouth it has been known
to restore the speech in paralytic cases, and made into
a conserve it is efficacious in scurvy and rheumatism. It
likewise increases the urinary secretion, and is good in the
gravel. But in whatever form it is used the root should be
fresh, for it loses the greater part of its efficacy in drying, and
becomes insipid.
In these more enlightened days it may possibly be difficult
to find persons with sufficient faith to try for themselves the
truth of the above remedies. Certainly, for my own part, I
should prefer, if suffering from rheumatism, a course of our
own thermal waters. I need hardly say that this plant has
ceased to be used in medical practice.'

1897: Herbal Simples Approved for Modern Use. William Fernie.
    This is a homeopathic based book which recommends a tincture
for the treatment of sore throat, swollen mucous membranes and vocal
soreness from shouting or singing, known as 'clergyman's sore throat'.
Unusually, Fernie also recommended Arum for 'an irresistible tendency to

------

*[1] A white waxy substance produced by the sperm whale, formerly used in
candles and ointments. It is present in a rounded organ in the head, where it
focuses acoustic signals and aids in the control of buoyancy. It was originally
believed to be whale sperm.*
*[2] Meaning the smallest unit of weight in the troy and avoirdupois systems,
equal to approximately 0.0648 grams and originally the weight equivalent to
that of a grain of wheat.*
*[3] A historical a unit of weight equal to 20 grains, used by apothecaries and
derived form the old French for a small pebble.*
[4] A Victorian reference to an aphrodisiacal effect here?

sleepiness, and heaviness after a full meal'. From five to ten drops of the tincture three times a day is recommended for an adult.

It is in his book *Herbal Simples* that Fernie reports on the use of Arum to treat neuralgia:

'Recently a patented drug, 'Tonga', has obtained considerable notoriety for curing obstinate neuralgia of the head and face - this turning out to be the dried scraped stem of an Aroid (or Arum) called *Raphidophora vitiensis*, belonging to the Fiji Islands. Acting on the knowledge of which fact some recent experimenters have tried the fresh juice expressed from our common *Arum maculatum* in a severe case of neuralgia which could be relieved previously only by Tonga: and it was found that this juice in doses of a teaspoonful gave similar relief.'

1931: Grieve's *A Modern Herbal*.

A modern and comprehensive book compiled in the style of the historical herbals with updated knowledge. Interestingly, Grieve introduces Arum's use as a diuretic and stimulant and describes its purgative nature, before rejecting its use for modern day herbalists, saying that it is no longer considered sufficiently safe to use.

She does still give its homeopathic uses (the same as stated by Fernie) as well as its usage for treating ringworm when its juice is mixed with lard and applied topically, though even here she qualifies it by sensibly stating that contact with the skin will raise a blister. An online version of the book is available at www.botanical.com

1947: Fournier's *Plantes Medicinales et Veneneuses de la France*.

Mentioned by Cecil Prime as the source of the child with the tumour on its arm who was cured by the application of Arum leaves and of the tale of a horse dying after having a fresh cut treated with an Arum leaf decoction, as per directions according to Dioscorides.

In this case, the horse owner must have misread Dioscorides, for the author does give a recipe for treating wounds with an 'Arum' but it is not the European Arum - it is *Dracunculus*, the most dramatic-looking of the European aroids, found around the Mediterranean area and, as Dioscorides says, it has a stalk 40 inches high! This was the Arum with which Dioscorides was most familiar, and it is definitely not to be mistaken for our more humble British Arum.

A CULINARY
TIMELINE

380–327 BCE: Theophrastus' *Enquiry into Plants*.
Theophrastus gives the first description of Arum being used as food: 'The root of cuckoo-pint is also edible, and so are the leaves, if they are first boiled down in vinegar; they are sweet, and are good for fractures'. He describes how 'men invert the roots of cuckoo-pint before it shoots, and so they become larger by being prevented from pushing through to make a shoot'. This way, all of the nourishment is drawn into the root and not wasted in the shoots and leaves.

65–75 BCE: Dioscorides' *De Materia Medica*.
He describes it as being suitable as a vegetable, with the leaves either being preserved in salt or boiled. The root was also edible, particularly when roasted with honey. Dioscorides also says that the leaves are good to eat and will preserve cheese if it is wrapped up in the leaves.

77 BCE: Pliny's Encyclopaedia.
Arum has a number of entries in Pliny's *Naturalis Historia*. Like Theophrastus, he mentions that it is known in Egypt as *aron* and that the root there is eaten raw (to wild-food eaters in the UK: please note that this is NOT the British Arum being referred to here!). Pliny also announces that already there is disagreement about whether Arum is one plant or many, on account of the different variations found around Europe, all quite similar but with noticeable differences. It is already known as *aron*, *dracunculus* and *dracontium*, and the Arum that Pliny mostly discusses is not the British Arum but the Egyptian, a plant now known as *Arum colocasia* or Taro. This is a notable example of plant observation that was somewhat forgotten in the later herbals, which lumped most arums together as one.

     As an article of food, however, later herbals give the preference to the female plant, the male plant being of a harder nature and more difficult to cook.

1500s:

The arrival of modern, continental starching techniques.

1526: The *Grete Herball*

'It hath greate vertue in the leaves, but moe in its rote, but
yet it hath more vertu in the knottes that be about the rote.
It is gathered and clouden in middes and dyed.'

1551:   William Turner: *A New Herball, wherein are Conteyneed the Names
of Herbes, Part 1.*

'The roote, sede and leaves of aron, have the same properties
of dragon. The roote is layd unto both members with
cowdunge: and it is laid up and kept as dragones rootes
are: and because the rootes are gentler, they are delyred of
many to be eaten in those countreis whereas the rootes of
coccowpynt are not so bytinge hote as they are in England
and in Germany.'

1558:   The Elizabethan era is said to have commenced.

1578: Henry Lyte: *A Newie Herbal.*

'They have the like power, when they are three or four times
boyled, until they have lost their acrimony or sharpness to be
afterwards eaten in meats, as Galen saith.'

1583:   Philip Stubbes: *An Anatomy of Abuses in England...*'.

Philip Stubbes writes his infamous pamphlet complaining about ruffs and
starching.

1603:   The end of the Elizabethan era.

1640:   John Parkinson's *Theatrum Botanicum.*

In John Parkinson's 1629 herbal, there are two recipes for *Arum
maculatum.* In one, small pieces of the root are mixed with lettuce and
endive, while in the other, the dried root is powdered and sprinkled over
meat. These recipes are recommended for the 'unbidden unwelcome guest
to a man's table' because 'it will so burne and pricke his mouthe that he
shall not be able either to eate a bit more or scarce to speak for paine'.

1710: William Salmon's *The English Herbal*
'It is prepared by boiling it till it is soft and all the acrimony pas'd off. It then nourishes and is good for food.'

1798: Bulliard's *Historie des Plantes Veneneuses et Suspects de la France*
Three children of a woodman in France become ill with Arum poisoning. Only one survives.

1796: The Royal Society advertises for a replacement for wheat starch.

1797: Mrs Gibbs submits her solution to the Royal Society.
Thus begins the explosion of an industry supplying Arum-based starch to the mainland.
1853: Last recorded use of Arum for starch-making on the Isle of Portland.
Recorded by an unknown writer and related by Ann Pratt in 1870. The last recorded individual was a single old woman, name unknown, who was in 1853 the last remaining individual to be making Arum starch on the Isle of Portand.

1861: *British Medical Journal.*
Three cases of children poisoned by eating Arum.

1870: Irish Famine.
Arum used to make flour for food.
1884: Thomas Fernie: *Plant lore, legends and lyrics*
'This flower, the Arum Maculatum, is the English Passion-flower: its berries are highly poisonous, and every part of the plant is acrid; yet the root contains a farinaceous substance, which, when properly prepared, and its acrid juice expressed, is good for food, and is indeed sold under the name of Portland Sago. Starch has been made from the root, and the French use it in compounding the cosmetic known as Cypress powder.'

1885: *Hardwicke's Science Gossip*
'Mr. Green first informs his readers that if they have been rash enough to taste the 'root,' an antidote will be found

either in milk butter, or oil. Writing still of the 'roots' he goes on to say: 'When dried they become farinaceous and insipid, in which case they might be used for food in case of necessity; and by boiling or baking would probably afford a mild and in being reduced to powder it loses much of its acrimony; and there is reason to suppose that the compound powder which takes its name from the plant, owes its virtues chiefly to the other ingredients. The pulvis ari compositus, or powder composed of Arum, is therefore discarded from the London dispensatory, and, instead of it, a conserve is inserted, made by beating half a pound of fresh root with a pound and a half of fine sugar.'

1919: Sturtevant's *Edible Plants of the World*.
Sturtevant described the uses of various arums found on his explorations, including tropical varieties. The common factor is the presence of starch and similar ways of either preparing the roots to render them edible or to

extract this starch for other purposes.

He mentions that arums were cultivated in Guernsey for their starch to

make arrowroot and that *Arum maculatum* is eaten in Albania and Slavonia

(now Croatia), where is it made into a type of bread.

1931: Grieve's *A Modern Herbal*
Reiterates that the tubers can be eaten when baked and have been used for starch production.

1939:   Last recorded presence of Arum mortars on Isle of Portland

1960:   Cecil Prime: *lords and Ladies*.
Prime doesn't look at Arum from a specifically 'culinary' perspective, but throughout the historical sections of his book does list the main food-related uses to which Arum has been put.

2006 onwards.
Modern wild-food enthusiasts.
The most prominent of these currently are Fergus Drennan and Marcus

Harrison.
Iraq and Kurdish tribes. Arum is still gathered and cooked with lemon juice or similar cooking acid to neutralise the calcium oxalate. Part of the traditional dish of *Kubba*.

BIBLIOGRAPHY

Albre, J., Quilichini, A. & Gibernau, M., 2003. Pollination ecology of *Arum italicum* (Araceae). *Botanical Journal of the Linnean Society*, 141(2), 205–214.

Allen, G. 1881. *The Evolutionist at Large*, Chatto & Windus. Available at: http://archive.org/details/evolutionistatl00allegoog [accessed 19 August, 2012].

Allen, G. 1897. *The Evolution of the Idea of God: An Inquiry into the Origins of Religion*, H. Holt and company. Available at: http://archive.org/details/evolutionideago00allegoog [accessed 19 August, 2012].

Anon., 1526. *The Grete Herball.* Translated from French by Peter Treveris: London.

Anon., 1833. *History of Vegetable Substances Used in the Arts, in Domestic Economy, and for the Food of Man, Boston,* Lilly, Wait, Colman and Holden. Available at: http://archive.org/details/historyofvegetab02bostiala [accessed 12 January, 2013].

Anon., 1838. Stone quarries and beyond. *The Penny Magazine.* Available at: http://quarriesandbeyond.org/articles_and_books/isle_of_portland.html [accessed 12 January, 2013].

Anon., 1861. *The Physicians of Myddvai; Meddygon Myddvai, or The Medical Practice of the Celebrated Riwallon and his Sons, of Myddvai, in Caermarthenshire, Physicians to Rhys Gryg, Lord of Dynevor and Ystrad*

Towy, about the Middle of the Thirteenth Century, Llandovery: D.J. Roderic, London, Longman & Co. Available at: http://archive.org/details/ physiciansofmydd00llan [accessed 12 January, 2013].

Anon., 2004. Arum maculatum *(for advanced foragers only)*. BushcraftUK. Available at: http://www.bushcraftuk.com/forum/showthread.php?t=3862 [accessed 12 January, 2013].

Anon., 2008. Herbal Medicine Diploma Course. A Brief History of Herbalism. Lesson 1.

Anon., 2012. Wihtburh [Withburga]. Wikipedia, the free encyclopedia. Available at: https://en.wikipedia.org/w/index.php?title=Wihtburh&oldid =530340470 [accessed 12 January, 2013].

Anon., Arum Maculatum – homeopathic remedy. *Homeopathy Pro*. Available at: http://www.homeotexts.com/arum-maculatum.htm [accessed 1 November 1, 2010].

Anon., Arum Maculatum L. Inchem. Available at: www.inchem.org/ documents/pims/plant/pim045fr.htm [accessed 12 January, 2013].

Anon., Elizabethan ruffs. Available at: http://www.elizabethan-era.org.uk/ elizabethan-ruffs.htm [accessed 12 January, 2013].

Anon., Etruscan herbal. Culture medicine. Available at: http:// culturemedicine.com/?m=201010 [accessed 12 January, 2013].

Anon., Herbarium of Apeleius Platonicus. Old & Sold. Available at: http:// www.oldandsold.com/articles31n/herbals-5.shtml [accessed 12 January, 2013].

Anon., Hidden mystery of the Unicorn Tapestries revealed. Winterspells: Life on the Magical Path. Available at: www.winterspells.com/3391/ hidden-mystery-of-the-unicorn-tapestries-revealed/ [accessed 12 January, 2013].

Anon., How Shakespeare has been made a lawyer. Shakespeare Law Library. Available at: http://www.sourcetext.com/lawlibrary/devecmon/02. htm [accessed 12 January, 2013].

Anon., In defence of oxalic acid. Dews World. Available at: http://www. dewsworld.com/InDefenseofOxalicAcid.html [accessed 12 January, 2013].

Anon., Old time remedies. Available at: http://www.oldtimeremedies. co.uk/labels/arum.html [accessed 12 January, 2013].

Anon., Robert Greene. Theatre database. Available at: http://www. theatredatabase.com/16th_century/robert_greene.html [accessed 12 January, 2013].

Anon, St Withburga of Dereham. *Encyclopedia Britannica*. Available at: http://www.britannia.com/bios/saints/withburga.html [accessed 12 January, 2013].

Anon., *The Complete Herbal*. Bibliomania. Available at: http://www. bibliomania.com/2/1/66/113/frameset.html [accessed 12 January, 2013].

Anon., The Herbal of Pseudo-Apuleius. From cave paintings to the Internet. Available at: http://www.historyofinformation.com/expanded. php?id=2580 [accessed 12 January, 2013].

Anon., The hunt of the unicorn. Wikipedia. Available at: https://

en.wikipedia.org/wiki/The_Hunt_of_the_Unicorn [accessed 12 January, 2013].

Anon., The life of Thomas Nashe (1567–1701). Available at: http://www. luminarium.org/renlit/nashebio.htm [accessed 12 January, 2013].

Anon., The unicorn in captivity. The Metropolitan Museum of Art. Available at: www.metmuseum.org/collections/search-the-collections/700 07568?img=3#fullscreen [accessed 12 January, 2013].

Anon., The village minstrel. John Clare Weblog: The Village Minstral. Available at: www.johnclare.blogspot.co.uk/2007/02/village-minstrel.html [accessed 12 January, 2013].

Anon., UK Flowers of Our Lady. Intro Mary Garden. Available at: http:// www.mgardens.org/UK-FOL.html#_jmp0_ [accessed 1 May, 2010].

Arber, A. 1912. *Herbals. Their Origin and Evolution: A Chapter in the History of Botany*. Cambridge, UK: University Press.

Assmann, J., 2001. *The Search for God in Ancient Egypt*. Ithaca, NY: Cornell University Press.

Auden, H., 1922. *Starch & Starch Products*. London: Sir Isaac Pitman & Sons, Ltd.

Ayse, E. & Ersin, O., Focusing on the ethnobotanical uses of plants in Mersin and Adana provinces (Turkey). *Journal of Ethnobiology and Ethnomedicine*, 1.

Baker, A.E., 1854. *Glossary of Northamptonshire Words and Phrases; with Examples of their Colloquial Use, and Illus. from Various Authors: to which are*

Added the Customs of the County. London: J.R. Smith. Available at: http://archive.org/details/glossaryofnortha02bakeuoft [accessed 12 January, 2013].

Barabe, D. *et al.*, 2004. On the presence of extracellular calcium oxalate crystals on the inflorescences of Araceae. *Botanical Journal of the Linnean Society*, 146(2), 181–190.

Bay, D., 1995. Thermogenesis in the aroids. *Aroideana*, 18, 32–39.

Bedalov, M. & Küpfer, P., 2006. Studies on the Genus Arum (Araceae). *Aroideana*, 29, 108–131.

Benson, Earl, Esq., W., 1797. *Transactions of the Society for the Encouragement of Arts, Manufactures and Commerce.*

Bermadinger-Stabentheiner, E. and Stabentheiner, A., 1995. Dynamics of thermogenesis and structure of epidermal tissues in inflorescences of *Arum maculatum*. *New Phytologist*, 131(1), 41–50.

Bierzychudek, P., 1982. Life histories and demography of shade-tolerant temperate forest herbs: a review. *New Phytologist*, 90(4), 757–776. Stable URL: http://www.jstor.org/stable/243183

Biltcliffe, G., 2009. *The Spirit of Portland: Revelations of a Sacred Isle*. Dorset: Roving Press Limited.

Bostock, J., Pliny the Elder, The Natural History. Available at: http://www.perseus.tufts.edu/hopper/text?doc=Perseus:text:1999.02.0137. [accessed 12 January, 2013].

Boyce, P., 1993. *The Genus Arum*. A Kew Magazine Monograph. London:

HMSO Publications.

Brown, D., 2000. *Aroids: Plants of the Arum Family, 2nd Edition*. Portland, OR: Timber Press.

Bulliard, P., 1798. *Historie des Plantes Veneneuses et Suspects de la France*. Paris: Dugour.

Cao, H. 2003. *The distribution of calcium oxalate crystals in genus Dieffenbachia Schott. and the relationship between environmental factors and crystal quantity and quality* (Doctoral dissertation, University of Florida). Available at: http://etd.fcla.edu/UF/UFE0001245/cao_h.pdf [accessed March 2013].

Chicheley, C.P., 1968. *A Manual of Plant Names*. London: George Allen & Unwin, Ltd.

Christy, M., 1863. The indifferent coiling of arum spathes. *Journal of Botany, British and Foreign*, 52, 1914. Stable URL: http://www.archive.org/details/journalofbotanyb52trim

Cicnjak, L., Huber, D., Roth, H.U., Ruff, R.L. & Vinovrski, Z., 1987. Food habits of brown bears in Plitvice Lakes National Park, Yugoslavia. In: *Bears: their Biology and Management, Vol.* 7, pp. 221–226. Stable URL: http://www.jstor.org/stable/3872628

Clare, J., 1821. *The Village Minstrel and Other Poems* [in two vols]. Stamford, Lincolnshire: Taylor & Hessey.

Cockayne, T.O., 1864. Anglo-Saxon leechdoms, wort cunning and starcraft of early England. Available at: http://archive.org/details/

leechdomswortcun02cock

Coles, W., 1657. *Adam in Eden*. London: Nathaniel Brooke.

Collins, M., 2000. *Medieval Herbals, the Illustrative Traditions*. London: The British Library and Toronto Press.

Colombo, M.L. *et al.*, 2009. Exposures and intoxications after herb-induced poisonings: a retrospective hospital-based study. *Journal of Pharmaceutical Science and Research*, 2(2), 123–136.

Cooke, M.C. & Taylor, J.E. *Hardwicke's Science Gossip* : an illustrated medium of interchange and gossip for students and lovers of nature. London: Robert Hardwicke. Available at: http://www.biodiversitylibrary.org/item/18256 [accessed 12 January, 2013].

Coté, G.G., 2009. Diversity and distribution of idioblasts producing calcium oxalate crystals in *Dieffenbachia seguine* (Araceae). *American Journal of Botany*, 96(7), 1245–1254. Available at: http://www.amjbot.org/content/96/7/1245 [accessed 12 January, 2013].

Cristhwaite, H. 2007. *A Fire Not Blown: Investigations of Sacral Electrical Roots in Ancient Languages of the Mediterranean Region*. Metron Publications. Princeton.

Croat, T.B., 1998. History and current status of systematic research with Araceae. *Aroideana*, 21, 126–145.

Culpeper, N., 1880. *Culpeper's Complete Herbal: Consisting of a Comprehensive Description of Nearly All Herbs with Their Medicinal Properties and Directions For Compounding the Medicines Extracted from Them.*

London: W. Foulsham & Co, Ltd. Available at: http://archive.org/details/
culpeperscomplet00culpuoft

Deane, G. Jack in the Pulpit, and Jill. Eat the Weeds and other things too.
Available at: http://www.eattheweeds.com/arisaema-triphyllum-jack-and-
jill-and-no-hill-2/ [accessed 12 January, 2013].

Dioscorides. *1 AD. De Materia Medica, a New English Translation*. [Transl.
Osbaldeston, T.A. & Wood, R.P.A. 2000]. Johannesburg, RSA: Ibidis Press.
Available at: http://www.cancerlynx.com/dioscorides.html

Dodoens, R., 1578. *A Niewe Herbal, or Historie of* Plantes. [Transl. Lyte,
H.].

Donohoe, T., 2002. Abstracts of the European Association of Poisons
Centres and Clinical Toxicologists XXII International Congress: 172. Few
symptoms reported following plant exposure in Ireland. *Clinical Toxicology*,
40(3), 374–375.

Dormer, K.J., 1960. The truth about pollination in Arum. *New Phytologist*,
59(3), 298–301.

Drennan, F., *Wild Man Wild Food – Fergus Drennan the Forager*. Available
at: http://www.wildmanwildfood.com/pages/wildfoodmonthdays1-31.html
[accessed 12 January, 2013].

Dunbar, A.B.C., 1904. *A Dictionary of Saintly Women*. London: G. Bell.
Available at: http://archive.org/details/saintlywomen01dunbuoft [accessed
12 January, 2013].

Dyer, T.F.Thisleton., 1889. *The Folk-lore of Plants*. Available at: http://www.

gutenberg.org/ebooks/10118 [accessed 12 January, 2013].

Ekrem, S., Turkish orchids and salep. *Acta Pharmaceutica Turcica*, 44, 151–157.

Eland, S., 2008 *Arum maculatum*. Available at: http://www.plantlives.com/plant_biogs.php [accessed May 2010].

Espíndola, Anahí, et al. 'New insights into the phylogenetics and biogeography of Arum (Araceae): unravelling its evolutionary history.' *Botanical Journal of the Linnean Society* 163.1 (2010): 14-32.

Available at: http://doc.rero.ch/record/20924/files/Espindola_Anahi_-_New_insights_into_the_phylogenetics_and_biogeography_20101109.pdf [accessed March 2013]

Fernie, W.T., 1897. *Herbal Simples Approved for Modern Uses of Cure*. Available at: http://www.gutenberg.org/ebooks/19352 [accessed 7 January, 2013].

Folkard, R., 1884. *Plant Lore, Legends, and Lyrics. Embracing the Myths, Traditions, Superstitions, and Folk-lore of the Plant Kingdom*. London: S. Low, Marston, Searle, and Rivington.

Fournier, J.A., 1947. *Plantes Medicinales et Veneneuses de la France*.

Franceschi, V. R. & Horner, H.T., 1980. Calcium oxalate crystals in plants. *The Botanical Review*, 46(4), 361–427.

Franceschi, V.R. & Nakata, P.A., 2005. Calcium oxalate in plants: Formation and function. *Annual Review of Plant Biology*, 56(1), 41–71.

Francisco Suárez de Ribera, D. Pedacio Dioscorides Anazarbeo, annotado por el doctor Andres Laguna ..., en la imprenta de Domingo Fernandez de Arrojo, 1733. Available at: http://archive.org/details/ pedaciodioscori00prgoog [accessed 19 August, 2012].

Frazer, W., 1861. Abstract of cases of poisoning by *Arum maculatum*. *British Medical Journal*, 1(25), 654–655. Available at: http://www.ncbi.nlm. nih.gov/pmc/articles/PMC2287751/ [accessed December 2011].

Freethy, R., 1985. *From Agar to Zenry; a book of plant uses, names and folklore*. Marlborough., Wilts: The Crowood Press.

Froberg, B., Ibrahim, D. & Furbee, R.B., 2007. Plant poisoning. *Emergency Medicine Clinics of North America*, 25(2), 375–433.

Frohne, D. & Pfänder, H.J., 2005. *Poisonous Plants: a Handbook for Doctors, Pharmacists, Toxicologists, Biologists and Veterinarians*. Portland, OR: Timber Press.

Fuchs, L., 1542. *De Historia Stirpium*,

Gardham, 2002. *Fuch's* [sic] *Great Herbal*. University of Glasgow Special Collections. Available at: http://special.lib.gla.ac.uk/exhibns/month/ oct2002.html [accessed 12 January, 2013].

George, U., 1974. *A Dictionary of Plants Used by Man*. London: Constable.

Gerard, J., 1597. *The Herball or Generall Historie of Plantes*. London: John Norton.

Gerard, J., 1899. Science and scientists : some papers on natural history. London: Catholic Truth Society. Available at: http://archive.org/details/ sciencescientist00gerauoft [accessed 19 August, 2012].

Gibernau, M. *et al.*, 2005. Physical constraints on temperature difference in some thermogenic aroid inflorescences. *Annals of Botany*, 96(1), 117–125.

Gibernau, M., Favre, C. *et al.*, 2004. Floral odor of *Arum italicum*. *Aroideana*, 27, 142–147.

Gibernau, M., Macquart, D. & Przetak, G., 2004. Pollination in the genus Arum–a review. *Aroideana*, 27, 148–166.

Givens, J.J.A., Reeds, K.M. & Touwaide, A., 2006. *Visualizing Medieval Medicine and Natural History: 1200–1550.* Farnham, Surrey: Ashgate Publishing, Ltd.

Gledhill, D., 2002. *The Names of Plants, 3rd edn*. Cambridge: Cambridge University Press.

Goodyer, J., 1933. *The Greek Herbal of Dioscorides. Illustrated by a Byzantine. Edited by* Robert Gunther.

Green, R., 1589. Preface. In: *Menaphon*. London: Sampson Clarke.

Greene, R., 1814. Preface to Greene's *Menaphon*. In: *Menaphon*. London: Longman.

Grieve, M., 1931. *A Modern Herbal: the medicinal, culinary, cosmetic and economic properties, cultivation and folk-lore of herbs, grasses, fungi, shrubs & trees with all their modern scientific uses*. London: Jonathan Cape.

Grigson, G., 1955. *The Englishman's Flora*. London: Phoenix House.

Grigson, G., 1973. *A Dictionary of English Plant Names*. Cambridge: Allen Lane.

Halevy, A.H., 1985. Arum. In: *CRC Handbook of Flowering, vol. VI*, pp. 52–74.

Harrison, A.P. & Bartels, E.M., 2006. A modern appraisal of ancient Etruscan herbal practices. *American Journal of Pharmacology and Toxicology*, 1(1), 21–24.

Harrison, A.P. & Turfa, J.M., 2010. Were natural forms of treatment for *Fasciola hepatica* available to the Etruscans? *International Journal of Medical Sciences*, 7(6), s16–s25.

Harrison, M., n.d. *Arum maculatum*/Cuckoopint or Lords-and-Ladies as a foodstuff. Country Lovers. Available at: http://www.countrylovers.co.uk/wfs/arum.htm [accessed 12 January, 2013].

Hedrick, U.P. (ed.), 1972 [1919]. *Sturtevant's Edible Plants of the World*. New York: Dover Publications.

Hill, J., 1756. *British Herbal: a history of plants and trees native to Britain, cultivated for use, or raised for beauty*. London: T. Osborne and J. Shipton.

Hort, Arthur, Sir. (ed.), 1916. *Theophrastus*.

Ito, K. *et al.*, 2011. Identification of a gene for pyruvate-insensitive mitochondrial alternative oxidase expressed in the thermogenic appendices in *Arum maculatum. Plant Physiology*, 157(4), 1721–1732.

Ivegard, T., 2009. *The Anglo-Saxon Charms*. Available at: www.wizros.com/SCA_Docs/Anglo-Saxon%20Charms.doc [accessed 1 December, 2010].

Janick, J., 2003. Herbals: the connection between horticulture and medicine. *HortTechnology*, 13(2), 229–238.

Kite, G.C. et al., 1998. Inflorescence odours and pollinators of Arum and Amorphophallus (Araceae). Reproductive biology. Kew, UK: Royal Botanic Gardens, pp.295–315.

Korth, K.L. *et al.*, 2006. *Medicago truncatula* mutants demonstrate the role of plant calcium oxalate crystals as an effective defense against chewing insects. *Plant Physiology*, 141(1), 188–195.

Kostman, T.A. *et al.*, 2001. Biosynthesis of L-ascorbic Acid and conversion of carbons 1 and 2 of L-ascorbic acid to oxalic acid occurs within individual calcium oxalate crystal idioblasts. *Plant Physiology*, 125(2), 634–640.

Kowalchik, C. & Hylton, W.H., 1998. *Rodale's Illustrated Encyclopedia of Herbs*. Emmaus, PA: Rodale.

Lack, A.J. & Diaz, A., 1991. The pollination of *Arum maculatum* L. – a historical review and new observations. *Watsonia*, 18, 333–342.

Lamarck, J.-B. de & Candolle, A.-P. de, 1805. *Flore française*, Chez H. Agasse.

Lewis, L.S., 2007 *Handbook of Poisonous and Injurious Plants*. New York: Springer.

Leyel, C.F. (ed.), 1992. *A Modern Herbal, 3rd edn*. London: Tiger.

Lim, K.T., 2012. *Edible Medicinal and Non-Medicinal Plants, Vol. 2*: Fruits. New York: Springer.

London, K., 1982. *The History of Birth Control. The Changing American Family: Historical and Comparative Perspectives*. Available at: http://www.

yale.edu/ynhti/curriculum/units/1982/6/82.06.03.x.html [accessed 6 June, 2012].

Loudon, J.C. (John C., A. & R. Spottiswoode, printer & United States. Patent Office. Scientific Library, former owner D., 1826. *The gardener's magazine and register of rural & domestic improvement*, London : Longman, Rees, Orme, Brown and Green. Available at: http://archive.org/details/gardenersmagazin03loudo [Accessed February 9, 2013].

Luis, L., 2011. *Arum maculatum*. Available at: http://www.fkog.uu.se/course/essays/arum_maculatum.pdf [accessed 12 January, 2013].

Lyly, J., 1601. *Love's Metamorphosis: A Witty and Courtly Pastorall*. Wood.

Lyly, J., 2012. *John Lyly. 18 Poems*. Available at: www.PoemHunter.com.

Lyly, J., Fairholt, F.W. (Frederick W. & Internet Archive (Firm)), 1858. *The Dramatic Works of John Lilly (the euphuist.)*. London: J.R. Smith. Available at: http://archive.org/details/dramaticworksofj02lyly [accessed 12 January, 2013].

Lyte, H., 1578. *A Niewe Herball or Historie of Plantes*. [Transl. Dodoens, R.] London: Edward Griffin. 1586 version available at: http://archive.org/details/mobot31753000811148 [accessed January 2013].

Mabey, R., 1996. *Flora Britannica, 1st edn*. London: Chatto & Windus/ Sinclair Stevenson.

Marc, G., Macquart, D. & Przetak, G., 2004. Pollination in the genus Arum – a review. *Aroideana*, 27, 148–166.

Mats, R., 1526. The English plant names in *The Grete Herball*: a contribution to the historical study of English plant-name usage. *Acta Universitatis Stockholmiensis, Stockholm Studies in English LXI.* Stockholm (1984): Almqvist & Wiksell International. SEK 92. Pp 110. ISBN 91-22-00710-5. ISSN 0346-6272.

McLaughlin, J., 2009. *A Word or Two about Gardening.* Orlando, FL: University of Florida.

Meddygon Myddvai, Unknown?, 1897. *Physicians of Myddvai.* [Transl. Pughe, J.].

Meehan, T., 1878. *The Native Flowers and Ferns of the United States in their Botanical, Horticultural and Popular Aspects.* Boston: L. Prang. Available at: http://archive.org/details/nativeflowersf01meeh [accessed 18 August, 2012].

Melamed, S., 2009. *Arum Dioscorides: an edible wild plant.* Available at: http://www.sarahmelamed.com/2009/02/wild-poisonous-plants-for-dinner/ [accessed 12 January, 2013].

Meyer, F. & Gustav, T., 1999. *The Great Herbal of Leonhart Fuchs: De Historia Stirpium, 1542.* Stanford, CA: Stanford University Press.

Midland Union of Natural History Societies, 1884. The Midland Naturalist : Journal of the Midland Union of Natural History Societies with which is incorporated the entire transactions of the Birmingham Natural History and Microscopical Society. London: D. Bogue. Available at: http://www.archive.org/stream/glossaryofnortha02bakeuoft/glossaryofnortha02bakeuoft_djvu.txt [accessed 12 January, 2013].

Murrell, W., 1884. *What to Do in Cases of Poisoning.* Available at: http://

archive.org/details/whattodoincases01murrgoog [accessed 12 January, 2013].

Myddvai, M. & Williams, J. ab I., 1861. *The Physicians of Myddvai: Meddygon Myddfai, or The medical practice of the celebrated Rhiwallon and his sons, of Myddvai, in Caermarthenshire, physicians to Rhys Gryg, Lord of Dynevor and Ystrad Towy, about the middle of the thirteenth century.* [From ancient mss. in the libraries of Jesus college, Oxford, Llanover, and Tonn; with an English translation; and the legend of the lady of Llyn y Van, D.J. Roderic.] London: Longman & Company.

Myer, F.G., 1999. *The Great Herbal of Leonhart Fuchs.* Stanford, CA: Stanford University Press.

Nash, T., 1589. *Preface to Menaphon, Green, A.M.* (reprinted 1814). London: Longman, Hurst, Rees, Orme & Brown.

Nazeen, W., 1861. Case of poisoning by *Arum maculatum. British Medical Journal.*

Nelson, L.S., Shih, R.D. & Balick, M., 2006. *Handbook of Poisonous and Injurious Plants, 2nd edn.* New York: Springer.

Newton, J., 1805. *A Complete Herbal.* London: J.D. Dewick/Allen & Co.

Nokes, R.S. *The Old English Charms and King Alfred's Court.* Available at: http://hompi.sogang.ac.kr/anthony/mesak/mes101/Nokes.htm [accessed 12 January, 2013].

Norfolk and Norwich Archaeological Society, 1852. Norfolk Archaeology. Available at: http://archive.org/details/norfolkarchaeol07socigoog

[accessed 12 January, 2013].

Norman, J. *From Cave Paintings to the Internet*. History of Information. Available at: http://www.historyofinformation.com [accessed 12 January, 2013].

O'Boyle, C., 1998. *The Art of Medicine. Medical Teaching University of Paris 1250–1400*. Leiden: Brill.

Pallas, P.-S., 1802–1803. *Travels through the Southern Provinces of the Russian Empire, 2 vols*. London: Longman and Kees.

Parkinson, J., 1640. *Theatrum Botanicum*.

Patil, J.D, 2007. *Group Study in Homeopathic Materia Medica*. New Delhi: B. Jain Publishers (P) Limited.

Pilegaard, K., 2009. Compendium of botanicals that have been reported to contain toxic, addictive, psychotropic or other substances of concern on request of EFSA. *The European Food Safety Journal*, 7.

Plowden, C.C., 1968. *A Manual of Plant Names*. London: George Allen & Unwin, Ltd.

Pollington, S., 2000. *Leechcraft, Early English Charms, Plantlore and Healing*. Norfolk: Anglo-Saxon Books. Ltd.

Porter, E.M., 1969. *Cambridgeshire Customs and Folklore*. London: Routledge & K. Paul.

Potterton, D. (ed.), 2002. *Culpepper's Colour Herbal*. Chippenham: Foulsham.

Pratt, A., 1855. *The Flowering Plants of Great Britain*. Society for

Promoting Christian Knowledge. Available at: http://books.google.co.uk/books?id=4i4OAAAAQAAJ [accessed 12 January, 2013].

Prime, C.T., 1960. *Lords & Ladies*. London: Collins.

Prychid, C.J., Jabaily, R.S. & Rudall, P.J., 2008. Cellular ultrastructure and crystal development in Amorphophallus (Araceae). *Annals of Botany*, 101(7), 983–995.

Rhind, W. *et al.*, 1855. *A History of the Vegetable Kingdom; embracing the physiology of plants, with their uses to man and the lower animals, and their application in the arts, manufactures, and domestic economy. Illustrated by several hundred figures*. London: Blackie and Son. Available at: http://archive.org/details/cu31924000606388 [accessed 12 January, 2013].

Rhizopoulou, S. & Katsarou, A., 2008. The plant material of medicine. *Advances in Natural and Applied Sciences*, 2(2), 94–98.

Riddle, J.M., 1992. *Contraception and Abortion from the Ancient World to the Renaissance*. Cambridge, MA: Harvard University Press.

Robertson, J. *Arum maculatum, Cuckoopint, Lords and Ladies. The Poison Garden*. Available at: http://www.thepoisongarden.co.uk/atoz/arum_maculatum.htm [accessed 12 January, 2013].

Rohde, E.S., 1922. *The Old English Herbals*. London: Longmans Green & Co. Available at: http://www.gutenberg.org/files/33654/33654-h/33654-h.htm [accessed January 2013].

Royal College of Physicians, 1829. *Flora Medica*. London: J. Davey.

Rubin, S., 1970. The medical practitioner in Anglo-Saxon England. *The*

Journal of the Royal College of General Practitioners, 20(97), 63.

Ryden, M., 1984. *The English Plant Names in the Grete Herball.* Stockholm: Almqvist & Qiksell International.

Saadi, S. & Mondal, A.K., 2011. Studies on the calcium oxalate crystals of some selected aroids (Araceae) in Eastern India. *Advances in Bioresearch*, 2, 134–143.

Sage-femme Collective, 2008. *Natural Liberty. Rediscovering Self-induced Abortion Methods.* Sage-Femme Collective. Nevada. Available at: http://www.scribd.com/doc/22321349/Natural-Liberty-Rediscovering-Self-Induced-Abortion-Methods (accessed 6 June, 2012). [For the Arum entry: http://www.scribd.com/doc/22321349/42/Arum (accessed 6 June, 2012)].

Sakai, W.S. & Hanson, M., 1974. Mature raphid and raphid idioblast structure in plants of the edible aroid genera *Colocasia*, *Alocasia*, and *Xanthosoma*. *Annals of Botany*, 38(3), 739–748.

Salmon, W., 1710. *The English Herbal.* London.

Skinner, C.M., 1911. *Myths and Legends of Flowers, Trees, Fruits, and Plants: in all ages and in all climes.* Philadelphia, PA: J.B. Lippincott Co. Available at: http://archive.org/details/mythslegendsoffl00skin [accessed 12 January, 2013].

Smith, B. & Meeuse, B., 1966. Production of volatile amines and skatole at anthesis in some arum lily species. *Plant Physiology*, 41, 343–347.

Sowter, F.A, 1949. Biological flora of the British Isles: *Arum maculatum. Journal of Ecology*, 37, 207–219. Available at: http://www.

britishecologicalsociety.org/journals_publications/journalofecology/
biologicalflora.php

Spratt, G., 1829. *Flora Medica, Vol. 1*. Oxford: Callow & Wilson,
Oxford University. Available at http://books.google.co.uk/
ebooks?id=om0FAAAAQAAJ (accessed 6 January, 2012).

Sprengel, K., 1807–1808. *Historia rei herbariae, 2 vols (2nd edn)*. Altenburg:
Geschichte der Botanik.

Stilo, A. *Crateuas*. Available at: http://penelope.uchicago.edu/~grout/
encyclopaedia_romana/aconite/crateuas.html [accessed 12 January, 2013].

Stubbes, P., 1853. *Anatomy of Abuses in England in Shakespeare's Youth*.
London: New Shakespeare Society.

Tabernaemontanus, 1590. *Neuwe Kreuterbuch*. Frankfurt: Eicones
Plantarum.

Talbot, C.H., 1965. Some notes on Anglo-Saxon medicine. *Medical History*,
9(2), 156.

Topsell, E., 1967. *The History of Four-Footed Beasts and Serpents and Insects,
Vol. 1*. London: Frank Cass & Co.

Turner, W. *et al.*, 1965. *Libellus de re herbaria, 1538: and the names of herbes,
1548*. [Facsimiles, with introductory matter, Ray Society].

Usher, G., 1974. *A Dictionary of Plants Used by Man*. London: Constable.

Walker, T., 1969. *Fox on a Barn Door. Poems 1963–4*. London: Jonathan
Cape.

Wang, H. *et al.*, 2005. Bradykinin produces pain hypersensitivity by potentiating spinal cord glutamatergic synaptic transmission. *Journal of Neuroscience*, 25(35), 7986–7992.

Warner, D., 2007. Arum. A plant surprise in hedge cleaning exercise. *Irish Examiner*. Available at: http://www.irishexaminer.com/opinion/columnists/ dick-warner/arum-a-plant-surprise-in-hedge-cleaning-exercise-41500. html [accessed 12 January, 2013].

Warren, J.W., 1940. *Island and Royal Manor Of Portland: Notes on its History, Court Leet and Court of the Manor*. Available at: http://books.google.co.uk/ books?id=mfp9uAAACAAJ. [accessed 12 January, 2013].

Watson, J.T. *et al.*, 2005. Outbreak of food-borne illness associated with plant material containing raphides. *Clinical Toxicology*, 43(1), 17–21.

Watts, D.C., 2007. *Dictionary of Plant Lore*. London: Academic Press.

Weil, M.S., 1972. *Magiferous Plants in Medieval English Herbalism*. University of Kansas.

Wink, Michael. Mode of action and toxicology of plant toxins and poisonous plants. *Julius-Kühn-Archiv* 421 (2010): S-93.

Woodward, M., 1974. *Gerard's Herball: the Essence Therefor Distilled*. [From the edition of Johnson, T.H. (1636)]. London: Minerva.

World Carrot Museum. *The Ancient Herbals & Carrots*. Available at: http:// www.carrotmuseum.co.uk/herbalists.html [accessed 12 January, 2013].

Wright, C.E. (ed.), 1955. *Bald's Leechbook* [British Museum Royal Manuscript]. Copenhagen: Rosenkilde & Bagger.

Xu X, *et al.*, 1998. Association of petrochemical exposure with spontaneous abortion. *Occupational and Environmental Medicine*, 55(1), 31–36. Available at http://www.ncbi.nlm.nih.gov/pubmed/9536160 (accessed December, 2011).

www.ingramcontent.com/pod-product-compliance
Lightning Source LLC
Chambersburg PA
CBHW040952170526
45159CB00013B/3103